THINKING SYSTEMS

T0206650

"*Stafford Beer commented 'I wish people would read my books,' I do hope people will read this one. Rooted in the timeline of his developing thinking, Robin Asby introduces and explains core concepts in systems thinking, cybernetics and, in particular the Viable System Model to great effect. He elaborates that model, its conception and rationale elegantly and with great clarity.*

The ideas of this book are challenging to conventional thinking, the language as complex as is necessary but not more so, and the philosophical shift required to realise the benefits is well indicated. For those new to the subject Robin offers an accessible, clearly articulated and well executed introduction to thinking in the field. For those familiar with the core material he offers a range of insights which create scope for conversation and the potential to provoke discussion and debate. It is rare to find such scope in one book!

This work deserves shelf space from all those serious about exploring the world through the lens of cybernetics."

Dr. John Beckford, President, The Cybernetics Society and Visiting Professor at University College London and Loughborough University, UK

"*Three features stand out in this thoughtful and insightful journeying of an experienced systems thinking practitioner. Firstly,* **Thinking Systems** *is a very timely further contribution to what some might call the more hopeful 'post-normal' constructive and corrective stages of our otherwise destructive unsustainable trajectory of the Anthropocene Epoch. The book is part of a zeitgeist of systemic sensibilities reaffirming the impoverished nature of Western rituals of what might be called analycentric (WEIRD, reductionist, mechanistic etc.) ways of thinking and practice that are clearly inadequate and/or insufficient for dealing with our increasingly interconnected and complex World...*

Secondly, the systems literacy contribution in **Thinking Systems** *provides an invaluable retrieval of, and primacy for, the cybernetics tradition of systems thinking – and in particular ideas from Ross Ashby (law of requisite variety) and ensuing modelling from Stafford Beer (viable system model). It seems to me that in our enthusiasm for a rather restricted sense of systems thinking, as exemplified by 'getting the bigger picture' and/or system dynamics, we sometimes lose sight of pearls of wisdom associated with the language of cybernetics. Robin Asby draws on his own long experience as a systems thinking practitioner in marshalling these ideas of cybernetics alongside ideas and tools from others including Peter Checkland (soft system methodology) in proposing an alternative, more systemic mode of thinking in practice.*

My final take-away from **Thinking Systems** *is the reminder that the most challenging aspect of developing systems thinking in practice capabilities lies with engaging multiple perspectives. The final chapter in Part 1, 'Recognising the Importance of Perspective', gives primacy to this challenge, and particularly invites us to view systems as conceptual devices. The subsequent chapters (8 and 9) apply the language and literacy of systems thinking developed in Part 1 to two different areas of practice – governance and quantum systems. These two chapters demonstrate some rigour (resonance and relevance, as well as reliability) of the model and language developed in Part 1. A further and perhaps more significant measure of success however might be gauged from ensuing effectiveness of conversations with actual practitioners associated with these two fields (policy makers/ social science advisors, and natural scientists). If this publication can help make a shift from current normalised frameworks of engagement to post-normal framings, then there is indeed hope with possibly realising the sub-title promise of contributing towards a much needed organic language of harmony."*

Martin Reynolds, Qualification Lead for Postgraduate Programme in Systems Thinking in Practice at The Open University, UK and lead Editor of *Systems Approaches to Making Change: A Practical Guide* (2020, 2nd Ed.)

"A really good read, that makes what at first glance appears to be a complex subject easy to comprehend and intuitive and changes the way we see the world and its problems.

Robin uses his extensive knowledge of Systems Thinking, gained through a lifetime of work in the discipline and close contact with Stafford Beer, to change our perspective on the way we think. Using concepts and examples from a diverse range of subjects – from Quantum Physics, Neuroscience, Psychology, Biology, Complexity Science to his own life experiences – Robin takes us on a journey to help discover the value of thinking about our lives and everyday living from the perspective of the systematic relationships within it. He ties in many sources of thinking and examples from philosophy giving a comprehensive background in order to build our understanding throughout the book and then shows us how we can make practical use of this way of thinking in our democracies or by applying it to Quantum Mechanics. Highly recommended."

Captain Jonathan Huxley, PhD (University of Portsmouth – focused on Systems Thinking Cybernetics Social Psychology)

THINKING SYSTEMS

An Organic Language of Harmony for Human Survival

Robin Asby

Triarchy Press

Published in this first edition in 2021 by:
Triarchy Press
Axminster, England

info@triarchypress.net
www.triarchypress.net

Copyright © Robin Asby, 2021

The right of Robin Asby to be identified as the author of this work has been
asserted by him in accordance with the Copyright, Designs and Patents Act,
1988.

No part of this publication may be reproduced, stored in a retrieval system
or transmitted in any form or by any means including photocopying,
electronic, mechanical, recording or otherwise, without the prior written
permission of the publisher.
All rights reserved

ISBN: 978-1-913743-32-1
ePub ISBN: 978-1-913743-33-8
PDF ISBN: 978-1-913743-34-5

To the memory of Ceri Asby

Acknowledgements

Many people have contributed to the development of my thinking and the models that I describe in this book. I feel particularly indebted to Ernest Hutten, Emil Wolf, Len Mandel, and Jerry Meek, for the many conversations and discussions that I had with them. Most of all I am indebted to Stafford Beer for his patience as I grappled with his Viable Systems Model and its implications over the twenty years of many extended conversations. Unfortunately now all these people are deceased.

I believe it to be self-evident that a teacher learns more from their students than they do from their teacher. Many students have contributed to the development of these models in their applications of Systems Thinking to their areas of interest. I thank them all.

I would also like to thank those who have read and critiqued the drafts that I have produced: Denis Adams, Jane Searles, Douglas Haynes, Roger Duck, Alex Hough, Ted Winfield, Jon Walker, Angela Espinosa, and Jonathan Huxley. Their questions and comments enabled me to clarify my thinking and hopefully I have produced a logical and clear path for my readers to follow.

I am immensely grateful to my editor at Triarchy Press, Andrew Carey, for his close reading enabling me to clarify many points in the argument, and to Nick Littlewood for his contribution to refining my diagrams.

But most of all I would like to thank my partner Penelope Marrington for her support over the many years that it has taken to get to this point. In truth she has contributed so much that it is not possible to disentangle the contributions made.

"We cannot solve our problems with the same thinking we used when we created them."

Albert Einstein

Contents

Preface

This is a work about the explanatory power of Systems Thinking, which I hope will enable those interested in Systems Thinking to understand that power. It is the result of more than forty years of effort to understand intriguing puzzles and it follows a path of exploration that that I have travelled in that time. Exploration has always been a driving force for me in my life. The excitement of not knowing what is round the next corner, or over the next hill, has always held a fascination. There is always something new, but occasionally something really surprising. Coming across an unexpected vista is always a rewarding, and uplifting experience for me both in my physical world and in my intellectual world.

Early in the 1980s I set out on a project to better understand Systems Thinking, which had already interested me previously as a young physicist. The project started as a result of my first meeting with Stafford Beer, whom I had sought out because I was looking for answers to questions of government which were puzzling me at that time. In that first meeting, finding out that I was a physicist by training, he made the assertion that Quantum Mechanics was a branch of Systems Thinking. This intrigued me, and started me on the journey described in this book. My explorations over the years following led me to the conclusion that Beer's approach to Systems Thinking was distinct, and of much greater power than generally realised. His approach was developed through spending time in India and studying Eastern philosophy during his Second World War army service. That understanding synthesised with his understanding of Western philosophy and later with the new science of cybernetics gave rise to his Viable System Model and his unique approach to Systems Thinking.

Arthur Koestler's book *The Act of Creation* explores the way in which syntheses of two different perspectives are the root of new insights and steps forward in scientific understanding, as well as the root of humour and art. This explains for me why Beer's approach was unique in the development of the ideas of Systems Thinking. What I didn't expect was that pursuing my understanding of Beer's approach would produce a model of learning that demonstrates this very process of how two different perspectives could produce new insights.

I

The second surprise was to find the connection to the process philosophy of Alfred North Whitehead. I found that not only was there a connection, but Beer's foundation was a much more usable approach to process understanding. The systemic techniques used in the development of the Viable System Model provide a basis for a much wider, process-based exploration of phenomena of great interest to science in any field where process analysis could be used to advantage. To use process-based analysis tools to explore phenomena which are processes seems common sense, but Western Science does not habitually do this.

Western thinking is doubly disabled: it insists on thinking in terms of static objects, and then categorising in terms of the attributes of these objects. Nature does not do categories and neither really does it do objects. The natural world is an evolutionary processual domain, in which for us as observers time is inherent, so static unchanging things are not the place to start. When we human beings model nature we must remember this. What I have tried to do in this book is to introduce a Western thinker to modelling in terms of process, using the foundation developed by Beer.

Thomas Kuhn writes that a young scientific enterprise starts out with many competing strands of thought, the proponents of each vying for the dominance that will make their approach the accepted way of thinking. Eventually one approach, being more successful in its explanatory power, becomes the dominant paradigm. In historical terms Systems Thinking is just such a young scientific enterprise. The first ideas appeared in Russia in the early 1900s and in the West in the 1930s. It developed quickly during the Second World War, through to the 1970s. Further development has taken place over the last twenty years, particularly in the realm of management understanding and ecology. There are now competing strands exactly in the way Kuhn describes, the proponents of each strand seeking to persuade us of the merits of their version. After nearly forty years of my own exploration, it seems to me that Beer's approach has all the features to make it that dominant paradigm, within which the relationship to other approaches and between other approaches can be understood. I argue this because of my own experience in applying this approach, and also because of the successes I have had in my experience in teaching this approach and helping others to apply it to many different problem areas.

In Part 1, the first seven chapters of the book, I describe this journey, leaving out the blind valleys and the false starts, in a way that I hope can be understood by those interested in the potential of Systems Thinking as a powerful approach to understanding the natural world in which we live. In Part 2 I describe outlines of the results of applying the thinking to the two areas which got me involved in this project: governing, and Quantum Mechanics. These two areas are far apart in the academic world but in each case surprising insights result from this systemic approach.

It is clear from the state of the world that things are amiss in the arena of government. Fault lines have been highlighted by the events of 2007-2008, and those now of 2020. But for me discovering in 1969 the work of Rachel Carson and Jay Forrester, and later that of the Club of Rome and others on the probable evolution of our stewardship of planet earth, was an experience which changed my outlook. 50 years later the warnings contained in those works about our probable disastrous evolutionary path now looks highly likely to be correct, and time to change is short.

One real surprise is that I have yet to find in mainstream writing on government the word '*cybernetics*' and yet the comparison to steering a ship, the Greek origin of the word, appears in the writing of Plato on governing, and the word itself in the writing of Andre Marie Ampère on governing. Norbert Wiener defined cybernetics to be the discipline of "control in the animal and machine". Governing is a cybernetic study so perhaps if we take a look at governing through the lens of cybernetics the new thinking might help us to achieve change. This is the subject of Chapter 8.

The situation in the realm of Quantum Mechanics is also problematic. Quantum Mechanics began its development around the same time as Systems Thinking, in the late 1890s and early 1900s. Whilst there is now an accepted mathematical paradigm, there are many proposed interpretations of what those mathematical models actually mean. Again, there are a number of competing paradigms in exactly the way Kuhn describes. My own interest was sparked by the fact that the derivation of the equations governing the area of electrodynamics is a seriously flawed process, but one which gives rise to models that accord with experiment to a remarkable extent. As before, different paradigms have been proposed to interpret the results of these equations. But there has not been a systematic exploration of the systemic approach in the way that I have undertaken it and report in Chapter 9.

In both these cases this systemic approach shows promise and there seems much more to explore. But the journey described in this book ends, standing on a ridge waiting for the mists to clear to see what this new vista contains; maybe nothing, but perhaps more surprises.

Part 1: The Journey

1. Sketching the Route

Holistic thought involves an orientation to the context or field as a whole, including attention to relationships between a focal object and the field, and a preference for explaining and predicting events on the basis of such relationships. Analytic thought involves a detachment of objects from contexts, a tendency to focus on objects' attributes, and a preference for using categorical rules to explain and predict behaviour. This distinction between habits of thought rests on a theoretical partition between two reasoning systems. One system is associative, and its computations reflect similarity and contiguity (i.e., whether two stimuli share perceptual resemblances and co-occur in time); the other system relies on abstract, symbolic representational systems, and its computations reflect a rule-based structure.

(Henrich et al 2010)

(WEIRD - Western, Educated, Industrialised, Rich, Democratic)

1.1.   Introduction

I am writing about holistic thought but this is a difficult task because any writing results in a long linear string of sentences, paragraphs and chapters, whereas holism implies that all is connected. To overcome this problem I have envisaged the linear structure of this writing as an expanding spiral, illustrated in Figure 1.1. The first chapter is a sketch of the whole picture I propose, an exploratory sketch, the first time round the spiral. The rest of the book explores the same ground in much more detail, the second time round the spiral. Beyond that there is still much to be done in enlarging and exploring further, but this is left for later and for others to do.

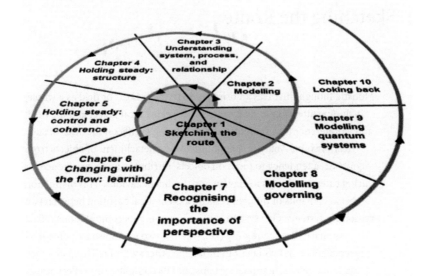

Figure 1.1 The path of exploration

This book is the result of a fifty-year personal journey of exploration and learning. As a young physicist I became puzzled by derivations of theoretical equations in which interconnections between different parts of a situation under consideration were disregarded. At around the same time I read Rachel Carson's *Silent Spring*, which alerted the world to the importance of hitherto disregarded interconnections between industrial waste products and environmental problems. As a result, I became interested in ecology and discovered the work of Jay Forrester, who was building computer models of the way in which the economic and natural worlds interact and evolve. This work became well known through the book *Limits to Growth* (Meadows et al) published in 1972. From this reading I developed an interest in the growing discipline of Systems Thinking which explored complex situations in which interconnections were important. In the early 1980s I came across Stafford Beer's approach to Systems Thinking and his Viable System Model, explained in several books – *Decision and Control, Brain of the Firm, Platform for Change,* and *Heart of Enterprise.*

The Viable System Model proved to be very important in the development of my own understanding set out here. I was alerted to his writing by a magazine article that described Project Cybersyn (see for example Beer 1972/1981, Medina 2011) which he had undertaken in Chile. The project involved three of my interests, government, computing and Systems Thinking. In the first meeting I had with Beer in the early 1980s he made the conjecture that Quantum Mechanics, a branch of physics, was very different from the rest of physics because it was a branch of Systems Thinking – Systems Thinking being holistic and different from the traditional norms of science. I found this assertion both surprising and fascinating because it connected the two apparently separate strands of my own thinking about government, and physics, and it set me firmly on the journey which has resulted in this book.

What I have tried to do here is to lay out my current thinking in a logical and coherent way, slowly building the layers of complexity. My aim is to describe my current understanding of how interconnections can be taken into account in any situation where they are important. I believe that this approach is particularly important in the understanding of the biological, ecological, psychological and sociological worlds, but important too in the world of government and physics.

1.2.   The Foundational Strands

The quotation at the start of this chapter draws the distinction between two ways of thinking. The paper it comes from makes the argument that WEIRD thinking (Western, Educated, Industrialised, Rich, Democratic), whilst being the norm in WEIRD societies, really is weird in human history and even in the context of today's world. Others, for example Richard Nisbett in his book *The Geography of Thought*, have also observed the differences between WEIRD societies and others. Jared Diamond (2013) also takes up the issue. WEIRD thinking is the basis of Western thought and Western science and the foundation of what we have called progress, but perhaps also, as Rachel Carson and many following her think, the underpinning of all that is wrong in our relationships within our current world.

I write this book as an exploration for my fellow WEIRD thinkers of holistic thinking, the first of the ways of thinking described in the quotation above. If we thought in this way, what difference would it make?

The foundation of the Western mode of thinking, and its underlying philosophy, is often attributed to Parmenides, a Greek philosopher who lived c515-c450 BCE. The starting point of his philosophy was that underlying all is a universal constant. Taking this as a foundation for thinking about thinking gives rise to the question: if I understand the world around as constant, how is the flux and change achieved that I see around me? This question has been thoroughly explored in Western philosophy, mathematics and science. However, the increasing complexity associated with problems in ecology, psychology, psychiatry, management and neuroscience, where interconnections are important, has challenged the adequacy of that understanding.

A generation earlier than Parmenides, another philosopher proceeded from what could be considered the opposite starting point. Heraclitus (c540-c480 BCE) said that "all is flux and change". Unfortunately, most of his writing is lost, but we do know from references to his writings which have come down to us that the writing of Heraclitus was considered very important in his own and later times. This foundation, that all is flux and change, is for me the foundation of the discipline of Systems Thinking. This is not always clear, even in introductory texts describing Systems Thinking, but I hope that in this writing I can clarify any confusion that the reader might have. Many writers share the belief that Systems Thinking cuts across the traditional academic disciplines and has arisen to explore complex problem areas where traditional approaches fail. I will return to this in the next section.

Two more ancient Greek philosophers are important here: Plato (c.427-347 BCE) whose so called 'Laws of Form' still puzzle us today (more of that in Chapter 2), and Aristotle (384-322 BCE) – one of the first philosophers to explore Plato's ideas. Aristotle considered that there were three important strands to the way he thought about things. Firstly, it is possible to describe an object, for example a house, by its purpose, what it is for; secondly, to describe it by its attributes, its characteristics; thirdly, one can say that both purpose and attributes are needed to understand something fully.

> ...[For] a house; the rationale will be something like 'a
> covering preventative of destruction by wind, rain, and sun'.
> But while one philosopher will say that a house is composed of
> stones, bricks, and beams, another will say that it is the form in
> these things for the given purposes. Who then is the natural
> philosopher among these? Is he the one who defines a house in
> terms of its matter and knows nothing of its rationale, or the
> one who defines it only in terms of its rationale? Or is he the
> one who defines it on the basis of both?
>
> Aristotle, *De Anima* (1986)

In the first way the object is described in terms of a relationship, relationship to me, some other person, or some other object. The problem with this is that different people have different relationships to the same object. Your neighbours' relationship to their dog will be different from your relationship to that dog. Not only that, but a single object can be the subject of different relationships to the same person. I may relate to a screwdriver using it as a lever to prise things apart or using it to screw a screw to fix things together. In all, this seems complex and slippery. At first glance and in contrast, the second way of describing an object in terms of its attributes contains no ambiguities; we each understand the same thing from the description: the neighbour's dog – small, long-haired and black; the screwdriver – yellow-handled, short, and stubby. So the second way seems more straightforward, simpler, and more 'constant', and is the way we in the West usually think.

This is not to say that in the West we have completely forgotten the first of these ways of thinking. The sciences of Semiotics and Biosemiotics, concerned with the interpretation of signs, take exactly the first approach. There has always been a concern about the way in which signs are interpreted (see, for example, John Deely's *Four Ages of Understanding*), and of course a sign is interpreted in relation to the interpreter. The same object may mean different things to different people, and in the natural world the same object may have different meanings for different species.

Research shows that there is a spectrum of different ways of thinking. I want to show that to get a grip on complex problem areas it is essential to start from the foundation of Heraclitus, rather than that of Parmenides. And, starting from there, there are strands of modern thinking about thinking that can be brought together to create a coherent and consistent philosophy that provides a language that can be used in complex situations more effectively than our WEIRD way of describing the natural world. This is not to say that our WEIRD way of thinking should be abandoned but that we should be using both appropriately.

Driven by the enormous success of Western thinking as a foundation for scientific advance, philosophical attempts have been made to explain holistic approaches in terms of WEIRD thinking – e.g. Nagel (1961). I turn that argument on its head. In hindsight it is not unexpected that the basic and 'natural' way of thinking is that practised throughout history and most of the world. And also, as might be expected, the Western way of thinking arises as a sophistication which enables flexibility but pays a price by losing precision and losing sight of the dynamic interconnectedness of the world around us. This I will explore further in Chapter 3. It is not surprising that if I am trained to think in terms of flexibility and simplicity it becomes very difficult to get a grip on complex interconnected problems.

1.3.   The Development of Systems Thinking

Systems Thinking as a recognisable discipline appeared in earnest in the 1940s and 1950s[1] with the recognition that complex systems often behaved in ways that were unexpected. In these situations, the interrelationships and interdependencies between the parts were as important in determining the dynamics as the behaviour and properties of the parts themselves – if not more important. The problem was that the mathematical modelling of relationships is difficult. This is exhibited in the science of cybernetics, one branch of Systems Thinking, which Norbert Weiner (1961) described as the science of "communication and control in the animal and the machine". This particular branch of Systems Thinking came from the recognition that mathematical modelling of control systems in different disciplines had produced the same understanding. In the late 1940s and early 1950s a series of conferences (the Macy conferences) was held to explore these issues.

This early work in Systems Thinking was based in the traditions of scientific enquiry fully within the normal WEIRD approaches but, when the ideas began to be applied to social situations, major difficulties were met. These difficulties destroyed the initial enthusiastic wave of progress. But in the 1970s and 1980s strands within the discipline of Systems Thinking abandoned the notion of scientific 'objectivity' as "the delusion that you can have an observation without an observer", words attributed to Heinz Von Foerster, a physicist and Systems Thinker. The social science interpretivist understanding that different observers of the same situation understand that situation in different ways came to be a fundamental tenet of Systems Thinking. In a complex situation, there are always multiple perspectives; understanding is relationship-dependent.

The second major departure from WEIRD thinking was the recognition that, in a complex situation where everything is connected to everything else, a situation, a system, cannot be easily differentiated out for study. The establishment of the boundary between that which is to be studied and that which is outside the study area is a judgement to be made by those who undertake the study, and again is relationship-dependent. These boundary judgements are crucial to developing understanding and so Systems Thinking draws also on this critical social theory tradition.

The quotation above from Aristotle sets out the way in which situations and experience are normally viewed in the Western world, in terms of objects

---

[1] Although Bogdanov's *Tektology* was published in Russian around 1913 and translated into German around 1920 (Biggart et al. 1998).

described by attributes, or purpose, but there are other ways. One view stands out: that of Alfred North Whitehead, who developed a process philosophy based in the American Pragmatist tradition of C. S. Peirce, William James and John Dewey. The change from description in terms of 'object' to that described by 'process' turns out to be vital – an importance that was perhaps missed because Whitehead's conception is not easy to use as an explanatory tool. However, when I analyse the various developments within the discipline of Systems Thinking, two pieces of work stand out as process-based and much used as powerful explanatory tools. So I come to the basic step away from traditional WEIRD thinking which in my teaching experience many WEIRD thinkers find difficult. It is this:

> *Building an approach based not on objects described by attributes but on process described by purpose is the key that unlocks a new world to WEIRD thinkers.*

This is the path I shall take.

In the late 1960s there were two complementary developments in Systems Thinking in the field of management, Stafford Beer's *Viable System Model* (VSM) (1972, 1979, 1985) and Peter Checkland's *Soft System Methodology* (SSM) (1981). The VSM was proposed by Beer as an archetype fractal model for an organisation, which could serve as a design framework for a new organisation or a diagnostic framework for improving the effectiveness and efficiency of an existing organisation. SSM on the other hand was proposed as a way of learning about an organisational situation in order to come to agreed action to improve that organisational situation. VSM and SSM were distinctive because they were both approaches to complex situations which were independently founded firmly on a teleological/functional process basis. These approaches together take fully into account the three distinguishing features of complex situations – an awareness of boundary judgement, multiple interconnectedness, and the ramifications of different perspectives.

The distinguishing feature of these approaches is that they explicitly recognise that a process is a transformation characterised by its relationship to its context. This relationship consists of the inputs which flow into the process – information and/or materials and energy – and the outputs which flow from the process – again information and/or materials and energy. A process is always purposeful/purposive, the purpose lies in the eye of the observer/definer of the process. Fundamental to the exploration is the relationship of the observer to the observed situation. But equally important is the observation that the output from the process in most circumstances changes the context and thus the overall behaviour is that of a looped system. I will expand on these two approaches in later chapters.

Ernest Nagel (1961) defined four types of explanation which he labelled 1) The deductive model; 2) probabilistic explanations; 3) functional or teleological explanations; 4) genetic explanations. Western thinking and, in particular, Western scientific thinking is based firmly within the first and second of these and has either rejected the last two as invalid or, with Nagel, attempts to explain them as derived from the first two. But it seems to me, and I hope to convince you when you reach the end of this book, that the most straightforward approach is to reject both these assertions. Functional or teleological explanations are neither invalid nor can they be derived from deductive models. The use of both VSM and SSM in management and elsewhere shows already that there is tremendous power in these systemic techniques. I hope to convince you that in many other areas a process-based approach is equally powerful and can be seen as the basis of thought.

WEIRD thinking, as I said above, is a brilliant simplification which introduces flexibility but loses precision. In other words, WEIRD thinking derives very simply from holistic thinking, not as Nagel tried to show in a complex argument, the other way round. This will be explored further in Chapter 7.

1.4.   Modelling

The first step which I explore further in the next chapter is to develop the idea of an *internal model*; that is a model which is built into the structure of an animal or indeed any living form, including *Homo sapiens*. This is an idea that has gained traction in recent decades and is becoming generally accepted.

The underlying concept is that living entities at some level perceive and respond to signals from their environment in order to enhance their probability of survival. Their purpose is to survive. The response takes place following a match between a perceived signal pattern falling onto sensors built into the living form and a pattern of structure, an *internal model*, also built into the living entity. The internal model transforms that recognised input signal pattern into some kind of output which enhances the chance of survival.

Hence an internal model has the structural pattern of *purposeful input→transformation process→output*; it is a *system*.

Signals – some combination of patterns of light, sound, touch, heat, acceleration, magnetic and electric field – are received by living entities from their environment. If there is a match between the signal pattern and the internal model then this can result in passive acceptance of the signal or active response to the signal. This is the the difference between the signal being interpreted as data – a passive acceptance, or as information – producing an active response.

Changes in an internal model, that is learning, take place either genetically for the species through the mechanisms of evolution, or for an individual through changes in its internal structures. An obvious example of individual learning is an animal with a nervous system learning through changes in its nervous system structure.

The psychology literature and the Systems Thinking literature refer to *mental models* (e.g. Johnson-Laird 1983, Senge 1990). Modern approaches to neuroscience, for example that of Antonio Damasio (2006), describe mental images and the way in which they are held, formed and used within the vast structured neuronal network of a brain. The idea of an internal model as I will use it certainly includes mental models but is a wider category and includes all elements of the response systems of living entities. The purpose of an internal model in any living form is to enable the prediction of the state and configuration of the environment in the future relevant to the living form concerned, given the state and configuration of the animal's environment now, so that the living entity can by some means enhance its survivability.

The crucial step, which allows everything that follows, is to realise that in its fundamental form the internal model is a structure that transforms perceptions into change. In other words, an internal model is a purposeful/purposive system; it receives inputs and transforms those inputs by some process to an output or outputs. It can be a response system involving the whole living entity, a tiny part of a nervous system multiply-coupled to other systems, or anything in between. But the *system* is the fundamental modelling building block for the remainder of this book. This is the crucial difference between this analysis and most other analyses. In other analyses, the thing or object described by its attributes is the fundamental building block. In my view, much of the complexity of published work describing the natural world falls away with this simple but fundamental change.

1.5.   System and Relationship

Despite living in a world of flux and change, I observe that there are seeming constants in my flow of experience. I see flows of people and goods into and out of local businesses, but the business appears much the same from week to week. I perceive the purposeful input→process→output structure that I defined in the last section; the business is a *system*. However, 'system' is a word in common use but not in common use well defined. There is a great deal of confusion in the Systems Thinking world but I require this very particular definition for the idea of a system, that of *purposeful input→process→output structure*. In this ideal system, signals and/or materials flow across a boundary into a process which transforms those inputs into new signals and/or materials which then flow out across the boundary as outputs back into the environment of the system. The system has a very particular relationship with its environment defined by these inputs and outputs. This notion of a system is the building block that is the basis of my modelling. I will explore these ideas in more detail in Chapter 3, but from here on I will reserve the word 'purpose' to characterise a process or a system in discussion, and the word 'function' when needing to discuss the use of an object such as Aristotle's house.

Using the building block of 'system', as I have now defined it, I can build an interconnected structure: the outputs of one process are flows that are then taken in by other processes. The parts of this interconnected structure are all transformation processes which are all interconnected through their inputs and outputs. This enables me first and most importantly to move my focus from the parts, the processes, to the relationships between the parts, the flows. This perhaps is one of the most important aspects of Systems Thinking: when modelling a complex structure, the move to process-based thinking automatically brings to the forefront of my thinking the interrelationships between the process parts, and then secondly the dynamic nature of my whole understanding. Both of these vital aspects of a dynamic world are largely missing if the understanding is based on objects. Just as easily as building models from systemic parts I can analyse any given system I define into its systemic parts. Any system can be thought of as consisting of a set of two or more interacting sub-systems.

Thirdly, thinking about systems in this particular way enables more precision in thinking. For example, most WEIRD thinkers, I suspect, would normally think of the tiger in the zoo as the same animal as the tiger in its natural habitat. The focus of this deduction is on the shape, colour and form of the tiger, i.e. the tiger as thing or object characterised by its attributes. But the tiger in each of these cases is not the same system; if the focus is on the relationship between the tiger and its environment it is easily seen that there are great differences between the two situations. In the first case the tiger's role is to be confined as an attraction for people to the zoo, in the second it is to enact the role of top predator in the wild. The interrelations between the tiger and its environment, the flows of information and materials from the animal to its environment and from its environment to the animal, are very different in these two cases.

This example shows the third essential element, the idea of *purpose*, which is inherent in the use of the word 'role'. 'Purpose' is a word whose meaning I must also explore, in order to define it well enough to contribute a precise meaning to my developing thesis. In order to do this I must introduce another idea, the notion of *perspective*. When trying to develop a model of a complex situation, perspective matters. A system is always defined from the perspective of the person doing the defining and, when considering complex systems, differences in definition abound. What is the purpose of a railway system – to transport people and goods; to make a profit and provide return on investment; to provide jobs; to provide directors with a large income? Just as in the case of the tiger, each of these possibilities reflects different environmental relationships, subtly different railway systems. The demise of the 'Railtrack' company, responsible at one time for maintaining the UK railway infrastructure, showed in stark reality how these subtle differences really do matter; but more of this later.

1.6.   Holding Steady: Structure

In an environment of flux and change, the question that needs to be answered is: how is stability achieved? I seem to be the same today as I was last week, last year, even ten years ago. I seem to be holding steady and together pretty well despite a lifetime of working in many different situations, retiring, and spending three months walking across France and Spain having never done anything like that before. Plants and animals around me survive and my world has much that seems to me to be constant. But as a gardener I do see change. Environments do change, but living entities survive and through their lifetimes remain pretty much the same; how do they do this?

These next chapters are devoted to exploring how stability can be achieved in an environment of flux and change. This is where the science of cybernetics comes to the fore. As I mentioned earlier, the Macy Conferences founded the science of cybernetics. The question of holding steady in a changing environment is a very general question, so perhaps in retrospect it is not so surprising that it appeared in several different fields of study. I start with what is perhaps the simplest case, what mechanisms are necessary for a system to hold to a particular purpose despite challenges from a changing environment.

Later I will come to the questions arising in the more complex cases where purpose changes. Following that I explore learning – the ability to create new internal models.

To unravel this extremely complex area I begin in Chapter 4 by exploring the next ideas that I need in order to propose answers to this question of how stability is achieved by means of a *feedback control loop*: these are the concept of *system state* and the concept of *variety*. Everyone is aware that friends can be angry, sad, happy or relaxed. There are many words which describe the state of a person. If I now consider the person as a system then I have examples of the state of a system. If a friend is angry then I need to take this into account when we converse – there will be subtle differences in our relationship from state to state but the core of our relationship will be unchanged. This is an example of a general property of systems, a system relates to its environment through its inputs and outputs. However even a very simple system can exist in different states, thus changing, usually in minor ways, the way in which the system relates to its environment. The number of possible states of a system is what is defined as the *Variety of the system*. *Variety* here has a technical systemic meaning which fortunately more or less coincides with its usual meaning.

The concept of variety is extremely important when I come to think about holding a system steady to a particular state, or group of states. In a world of flux and change, in order to hold a system steady I will need to control or manage the system to achieve that particular end. W. Ross Ashby was a British psychiatrist who first demonstrated the importance of variety in understanding control. He showed that, in order to achieve sustainable control of a system, the variety of control actions by a controller must be at least equal to the variety of the disturbances which the system could be subject to. This is the *Law of Requisite Variety*. For example, the state of a motor vehicle on a changing road is described by its lateral position on the road and its speed along the road, so we need at least control systems to manage these two variables – the person driving needs the steering mechanism, and the accelerator-brake combination to arrive safely at their destination.

The second important concept I need is the *Conant-Ashby Theorem* (Conant & Ashby 1970) which follows from the Law of Requisite Variety and states essentially that the quality of control or management of a situation depends upon the quality of the model built into the controller. In slippery road conditions the number of potential states of the car system increases: normal driving does not include skids. Understanding how a car behaves in those conditions, having a superior, higher variety model of car behaviour and how to control it, makes for a safer and a more skilled driver.

With the concept of *variety* I can now explore the structure necessary to deal with the large *variety* of challenges that would have to be met if stability is to be maintained. Even in the simplified case of a system maintaining an unchanging purpose the necessary structure is far from simple, but the result of the exploration – a fractal structure – does enable progress towards understanding how stability in a world of flux and change can be achieved. A good example of a system with an unchanging purpose is a human organisation. A human organisation, a business, a government or a charity is usually established with a fixed purpose and so is indeed a simpler organisation than any living animal which changes purpose even from moment to moment. I will explore the example of a restaurant. A restaurant is a familiar form that comes in many guises, but a purpose can commonly be defined as follows: to give customers a sufficiently enjoyable eating experience to cause them to come again. Using this example I explore the way in which I can analyse the restaurant into a set of interacting sub-systems, and this further into an interacting set of sub-sub-systems, and so on. Whilst this is simple in principle, a complex set of interacting systems soon results from a detailed analysis.

1.7. Holding Steady: Control, Coherence and Adaptability

The enduring legacy of the original founders of cybernetics is the understanding of the essential requirements needed to control or manage any system. Three abilities are needed. First, I must be able to watch the system under control to perceive departures from the desired state. This implies that a sub-system of the control system to sense changes is required. Secondly, I need an adequate model of the situation in order to be able to compare the actual state sensed with the desired state that is necessary for the purpose to be achieved. Thirdly, I need the means to adjust the state of the system if the external challenges move it from the desired state. In the example above it is the driver's eyes, ears and inner ear proprioceptors which sense the state of the car system, the driver's brain which contains the model, and the steering, accelerator/brake combination which effects any changes required. This mechanism is the classic cybernetic *feedback control loop* which constrains the system being controlled or managed to a required state or set of states.

Therefore, to manage the relationship between any system and its environment I need these three sub-systems:

- a *sensor* that is able to detect the state of the relationship between system and environment
- a *comparator,* that contains a model of the relationship between the system and the desired state
- and an *actuator,* a means to adjust the current state to the desired state.

One important understanding brought out by the fact that I am now modelling in terms of a process structure is that the controller sits not alongside the controlled system but takes an overview, sitting on a logical meta-level; meta meaning outside or above. This relationship was first understood and explored by Stafford Beer in his 1959 book *Cybernetics and Management* and is explored in Chapters 3 and 4

Given that a human organisation or a living entity consists of different interrelating parts, a set of sub-systems, what mechanisms are necessary for the parts of the entity to remain together? This is the question of coherence. But further than that, not only must the parts remain together but the entity must also remain a coherent whole, a stronger requirement. So then what mechanisms are necessary for the entity to remain this coherent whole? This is the internal control problem: how are the relationships between the parts held stable in a changing environment so that the entity can survive?

For the whole to remain coherent, the parts must communicate with each other. The most common complaint that I heard in my years at work was that people do not receive the information they need to do their job effectively and efficiently. Organisations are not generally designed to facilitate communication but designing a process-based structure which does facilitate communication makes a difference. The success of Beer's VSM model in aiding the diagnosis and design has shown this. Even without a designed, process-based structure, more two-way communication leads to more coherence. The effect of social media in enhancing the ability of people to communicate has been profound. For example, I can communicate in real time with colleagues around the world to discuss the ideas I am putting forward in this book. On the other hand, organisms in nature are built from the bottom up by trial and error. Organisms come together perhaps by chance and if they are capable of holding together (that is, the interrelationships have the necessary form), a new, more complex organism comes into being.

Any structure in a dynamic world has to face the problem of maintaining internal stability. The result of not maintaining these relationships can be seen in the break-up and demise of countries and other human organisations, and in the death of animals. Parasites evolve to be coherent with their host. They live as part of their host at least until their host is no longer needed. These internal relationships must be constrained in exactly the same way that the relationship to the external environment of the whole system must be constrained. Indeed these are two aspects of the same problem since the environment of a sub-system includes both the other sub-systems and the system environment. The mechanisms of constraint have exactly the cybernetic form described above. However, for an organisation to achieve the form required, which includes the necessary constraints on all the sub-systems for the system to remain coherent and whole, is far from easy.

Envisaging a system as a multilayered fractal structure of sub-systems enables an understanding of how stability is achieved in a changing environment, and Ashby's Law is satisfied. Sub-systems, parts of an organisation, parts of a country, parts of an animal, parts of a system... each operates to its own semi-autonomous purpose within a framework. The framework constrains the variety of the operation of the sub-systems to that which is necessary to maintain coherence. Bees get on with their life with one queen until the colony grows too big and some find themselves outside the pheromonal influence of that queen. The constraint has gone and these bees then create a new queen for themselves and leave to form a new colony in a new hive.

With multi-level systems, the relationship between two adjacent levels is crucial: the relationship of system to sub-systems. The lower level sub-systems must operate within an established framework. In a human organisation the management of the whole can always improve the quality of its control models by involving and including the knowledge contained at the level of the parts – something which is not often done. In this situation the framework is more likely to hold if it contains the knowledge contribution of the sub-systems. Hence the control structure should be, and must be, fractal to be successful.

In contradistinction, it seems that it is part of the WEIRD culture to believe that management and government (management of a society) can be achieved by a central function, on just one level. A process analysis leads to the conclusion that control mechanisms are needed at all levels. Organisations and countries break up because a central government does not have the requisite variety to govern successfully. Ashby's Law of Requisite Variety cannot be satisfied, nor can the Conant-Ashby Theorem. The *variety of potential command* of the governing or managing system is inadequate and the whole will not be maintained. A centralised hierarchy can never have requisite variety – it must always rely on fear and intimidation, bribery, or just acquiescence by lower levels with decisions made by higher levels on the assumption (by both levels) that the higher level has the right to do this – many of us are happy not to protest if decisions don't accord with what seems sensible. This does reduce the variety that needs managing, but these approaches do reduce viability, and especially the possibility of surviving in a changing world, because eventually control breaks down. Hierarchies which try to reduce variety in this way are rigid, the internal control models cannot be changed and so eventually they expire or succumb to revolution.

Lastly in Chapter 5 I begin the exploration of the necessity to change purpose. Viability of a living form is improved by its being able to respond to short-term environmental changes, which could be as simple as seasonal changes or the appearance of a threat. All living forms face these possibilities but, for an animal, the need to move around to look for food or to avoid predators needs greater flexibility. Even in the short term a mobile life form changes its relationship to its environment often, even from moment to moment. So, in adapting to a changing environment, the next set of challenges is for the individual organisation or animal to be able to change its purpose – to change its relationship to the environment to meet a change in the external environment. What are the mechanisms necessary to enable a change in the internal models in use to achieve this?

Again Ross Ashby (1952) was here first in discussing the essentials of *homeostasis*. To be able to change purpose there must be embedded in a system's structure a range of possible models enabling the recognition of threats or opportunities and then enabling an appropriate response. In its simplest terms, the survivability of a living entity will be enhanced by being able to move towards food and away from predators. Therefore, the first systemic requirement of adaptability is to be able to detect patterns in the environment which require a change in the relationship between entity and environment that pertains at any time. To do this an additional control loop is required, a mechanism to recognise those situations, food or predator, and to put in place the appropriate model to deal with that particular situation. This is the *feedforward control loop*. Once the new purpose is in place the feedback control loop, explored in the previous section, keeps the relationship aligned to that which is required.

So far it is assumed that the models at the heart of the control system, part of the structure of the whole system under consideration, already exist, perhaps developed through evolutionary processes. But for an animal with an extended lifetime this will not be sufficient; to remain viable in a changing environment the ability to change internal models is a necessity. In the next section and Chapter 6 I explore the changing of internal models, i.e. learning.

1.8.   Holding Steady: Learning

Attempting to hold steady to a particular constant relationship in a changing environment will not ensure survival. Nature's solution to this problem was first and foremost Darwinian evolution: the ability to produce progeny that could be different from the parent/s. This is usually referred to as species adaptation but could also be considered species learning. By species learning I mean changes in internal models introduced by the genetic and epigenetic processes in the transmission of models from generation to generation. As an environment changes, changes in internal models which improve the probability of successful reproduction and survivability, enable adaptation. But I will leave this to one side to consider only how individual living forms adapt to a changing environment.

Models are developed and maintained in the brain and nervous system. This consists of interconnected neurons and other supporting cells. A neuron is a system, it receives inputs from its environment and transforms those inputs into outputs to its environment. Interconnected neurons form a complex interconnected web of systems. Signals flow through this web of interconnected cells. Hebb (1949) proposed that the strength of inter-neuronal connections evolves. Edelman (1987) describes how at first, for any animal, this is driven by processes of maturation, but subsequently is driven by the animal's experience of its particular eco-niche. Interconnections that are *not* used, decay, and those that *are*, are strengthened. Furthermore, new interconnections come into being, guided by those signal flows (Lettvin et al 1959). The neurons are arranged in neuronal groups of varying sizes (Edelman 1987), in such a way that there is redundancy of function and also degeneracy of function.

This structure was recognised by Paul Cilliers (1998) to be a complex system – exactly as envisaged in the science of complexity theory and needed if we are to use the modelling techniques developed here. So in Chapter 6 I use the modelling techniques of the previous chapters to explore the idea that a model is encoded in the collective action of neurons: that is, the structure of the neuronal network determines the relationship between input and output.

Two further levels of complexity in an individual's control system are important in holding steady in a changing world. First, it is necessary to be able to hone recognition and response. Many years ago I was troubled by the fact that I could not identify most of the birds that visited my garden. I just saw birds. Where there was one model, I needed many. That one system, which gave me the understanding of 'bird', needed to be developed into many sub-systems. With the aid of pictures and descriptions and much practice I developed the ability to identify those birds. I continue in my efforts to learn to play the piano, slowly increasing my ability to hit the right notes, and hear that I can do that with increasing speed. Practice makes perfect. For a frog, an amphibian, it is sufficient to jump far at random when a potential predator approaches for it is of no consequence if it lands in water or on land. Hence evolution provides it with the means to jump far. But being able to choose a direction and distance to jump, to choose a response taking into account the terrain around, increases the chance of surviving.

Secondly, it is necessary to develop responses to new challenges, those that have never been met before. We have the ability to extract common patterns from different experiences, and then use that extracted model to evaluate and react to new experiences. This is a technique that is often used in teaching: I have both experienced and used it, introducing particular examples and then describing the common pattern. In my quest to be able to identify birds I can begin by identifying the commonalities of looks and behaviour between a buzzard and a red kite and thereby increase my ability to identify another bird of prey. Certainly, beginning to play the pieces that I practise on the piano competently enables me to play pieces that I have not met with some initial competence. Being able to recognise a generic pattern is a powerful aid to learning.

I would propose that these two abilities are rather simple in systemic terms. In the first case I am developing the ability to differentiate between two systemic processes: a recognition and response system is refined into two or more sub-systems. In the second case I identify a schema common to two recognition and response systems and then use that schema to guide reaction to a new situation. In systemic terms, a meta-systemic schema-system is developed from two systems, from which new systems can be developed using the first learning process. Both these possibilities increase the variety of responses available to environmental disturbance and therefore, according to the Law of Requisite Variety, increase the chance of survival. I will explore this further and the relationship between this approach and that of Jean Piaget (1954) in Chapters 6 and 7.

1.9.   An Underlying Philosophy

Science is a process, the process of learning about the environment in which we find ourselves. A scientist proposes and explores models that enable the prediction of how our environment might change depending on the actions we take. The models enable the engineering of substances, machines and constructions; they enable increases in energy usage and food production. The founder of the American Pragmatist tradition, C.S Peirce, stated that thought distinctions are meaningless unless they lead to a difference in practice. A scientist is a professional sceptic concerned with testing models as much as proposing them. Karl Popper formalised the testing process: once a new model is proposed, it is the scientist's obligation is to devise tests with the aim of testing the validity of that model. If the model survives the testing, then there is an obligation to broadcast the model for others to test.

Max Planck once famously said that science progresses one funeral at a time. As the proponents of a set of models pass away, new ideas gain acceptance. Thomas Kuhn explored this process, characterising the transition from one explanatory set of models to a different set of models with changed interpretation of a body of evidence as a change of paradigm. Inevitably, with an ever-increasing body of knowledge, subcultures form within the scientific milieu. A paradigm change takes time to propagate across subcultures.

In the West we act as if we believe that an exponentially expanding population, each of us expecting living standards to improve, can be sustained by extracting non-renewable resources on a finite planet. Since this belief is a clear piece of nonsense, if we are to sustain a human population without war, disease and starvation, the way we act must change. Therefore our beliefs, our models, must change. A new way of thinking, a new paradigm, is essential for a sustainable world. Thinking in terms of process, or systems, which is the focus of this book, enables modelling that brings out the interconnections missing from WEIRD thinking. The underlying philosophy is explored in Chapter 7.

1.10.   A Summary

No individual living form starts from scratch; all are born into the world with a starting set of inbuilt models. A human child starts its individual learning in the womb, then exposed to its econiche takes small steps adapting to and with the world around it. It inherits human form but adapts to the culture into which it was born, propagating both its genes and the culture.

The approach I propose is the result of more than seventy years of seeking explanations for the patterns I have experienced. I propose a new set of models derived from the many thought experiments I have carried out. Testing the models through thought experiments and against the models of other authors often led to an improvement in a model and some honing of the approach. For me they have also brought increased coherence to areas of understanding, so I now wish to communicate those models to others. It is for others then to test each model, in the same way as I have done, for verification. If more and more people accept the new models it becomes the accepted way of understanding the situations concerned. It becomes the 'objective reality' and replaces the previous objective reality.

Part Two outlines the two applications of the approach to the two areas which have driven my exploration, the problems of governing, and the problems of Quantum Mechanics. How do these modelling techniques answer the queries that started this long quest? I urge the reader to read on because in both cases the modelling does give answers. Governing is a dynamic process that is all about relationships so I contend that WEIRD thinking has little to offer but Systems Thinking has much to offer as described in Chapter 8. So too with Quantum Mechanics, Bernard d'Espagnat (2006) also concluded that Quantum Mechanics is a process theory and does not fit the realm of WEIRD thinking. A systemic approach to Quantum Mechanics as outlined in Chapter 9 might well be the answer to the many outstanding issues.

2. Modelling

2.1. Introduction

> Every living organism – a hunting lion, a grazing antelope, a stooping falcon
> – carries within itself a quantity of unique information. The information was
> acquired at great cost and hazard over millions of years in circumstances
> that can never be duplicated. Each set of data was scrutinised and tested
> against the most rigorous standards of proof. If the information finally
> proved sound, it was genetically incorporated in the animal's body, to be
> available for consultation across the succeeding millennia.
>
> To destroy a species is to lose the meticulously researched and proved
> information it contains. It is like plundering a library shelf and tearing up
> every handwritten volume labelled A-D. The loss is irreplaceable.
>
> (Nicholas Luard 1981)

Having sketched out in Chapter 1 the route I wish to follow, I now set out
along the second loop of the expanding spiral starting again at the beginning
of the journey. I suspect that any reader trained into WEIRD thinking, as I
was, would find some of Chapter 1 difficult to comprehend so on this loop of
the spiral I will explore in much more detail the ideas that I wish to convey.
But again I warn that writing about interconnected ideas is difficult so it may
be necessary on occasion to suspend judgement and simply read on.

2.2. The Idea of a Model

Like many boys of my generation I spent hours of my boyhood constructing
models built from a Meccano construction kit. I have many memories of the
cars, cranes, lorries and other working models that I constructed. At around
the age of eleven I began building flying model aeroplanes, with not much
success at first, but that then took up much of my time. Although not
consciously, I became naturally familiar with the idea that a model is some
thing, an entity, that is representative of some other entity out in the adult
world.

Although familiar with the idea of a model, it was only much later that I
became conscious that a model represents the original only in some respects.
The interesting question is this: what decides the respects that are represented

and those that are not? As a child I strove to achieve 'working' models, more than just something to look at: vehicles that could be steered, aeroplanes that flew.

In the adult world there are models designed just to be looked at. For example, an architect's model of a building is constructed to allow the people for whom the building is destined to have an impression of the shape and form of the building. It will not normally contain a representation of any of the working parts. The electrical wiring or the plumbing of the real building will not be present in the model. So the model represents only features of the real thing that are relevant for the purpose that the model is intended. Electrical wiring and plumbing are not part of what is required. However, the plans of the building are also a model, a representation of the building with a rather different purpose. The plumbing and electrical wiring will be shown on plans, but possibly on separate different plans intended for plumbers and electricians respectively. However, plans are still far from identifying everything about the eventual building. The purpose of the plans of a building is to have a model that will enable the builders to complete the construction of the building. It will usually assume that the builders know how to turn the plan into a building, so the plan doesn't look much like the building. Much is still omitted, but a skilled team of builders will not notice that, because they know exactly how to interpret the plan.

These few examples illustrate that a model is made for a purpose. A model is a model because its particular form and the structure of its parts match in some way the form and structure of the parts of the entity that is modelled. A model has some aspects of the real entity, those that are represented are present in the model, but not all aspects are present. It is the purpose of the model as decided by the modeller that decides the aspects of the entity that are represented in the model. Models are always made for a purpose, and whatever that purpose is decides what aspects of the real entity are represented in the model. These included aspects define the precise relationship between the model and the entity being modelled.

2.3.   Mental Models

Now I wish to examine another aspect of modelling because even though a modeller constructs a model with some purpose in mind, the observer of that model may not see the model in the same way that the modeller intended.

Investigations of perception find that the way we perceive the world around us depends on experience. Hermann Von Helmholtz (1821-1894) founded the modern idea of perception. Working in the area of optics and sight he proposed that what is perceived at any one time is a synthesis of the pattern of light falling on the retina of the eye and remembered patterns stored in the brain. Perception is not a simple reproduction of what is external. As a small child I remember being frightened by the faces that I saw in folded and crumpled cloth – curtains, sheets, and blankets – piled in my bedroom, to the extent of asking my mother to remove the offending items. Of course, there were no faces, but I saw faces. Richard Gregory (2009 p.198) puts it like this:

> Objects are attributed to images. I see a bowl of grapes on the
> table – as my brain attributes the images in my eyes to grapes,
> which I know from past experience, and they have a fair
> probability of being grapes.

There are many examples of optical illusions (Gregory 2009) that illustrate the brain interpreting externally perceived patterns in terms of remembered patterns. One example is a famous optical illusion first drawn by cartoonist W.E. Hill in 1915 and reprinted in the psychological literature by Boring (1930) reproduced at Figure 2.1. It is just a collection of black marks on paper, but your brain interprets this pattern of marks as – what? Do you see an old woman or a young woman? Both are possible, and your brain will pick one of these, and it may be quite difficult for you to change from this and see the other.

Figure 2.1: Black marks on paper?

This example illustrates the proposition that your brain contains some remembered pattern and then tries to interpret the pattern of black marks on the paper into a form that is familiar. That is, just as Helmholtz proposed, what you perceive is a synthesis of the light patterns falling on your eyes and your remembered patterns. Different people interpret the marks differently, some people see an old woman, some a young woman. With practice you can see both, but you can only see one at a time, so it seems that only one internal, remembered pattern can be used at any one time. In order for the pattern-matching to occur, enabling the interpretation of objects in the external world, these remembered patterns must contain in some way the structure and form of the objects in the external world. It is therefore natural to call them mental models.

I can think of my perception in terms of the use of my mental models, the use of the patterns that are stored in my brain to interpret the patterns of light, sound, touch, taste or smell that I receive from the outside world through a matching/recognition process. As far as the cartoon is concerned, I use my pre-existing brain model of a young or old woman to interpret the patterns of light and dark that my eyes capture from the marks on the page. Either I use a pre-existing model of an old woman or a pre-existing model of a young woman to interpret the signal received. I can use one or the other, and with practice switch between them. But whichever I use, what is actually on the paper is just a pattern of black marks; it is my brain that, using its remembered experience, is giving me the interpretation of a representation of an old or young woman; just as it was my brain producing the frightening images of patterns in the folds of the cloth that I saw as a child.

In summary then, I can define a *mental model* as just such a remembered pattern gathered from past experience of some external entity – grapes on a table, a young or old woman – stored within a brain or nervous system of a human being, or for that matter any other brain or nervous system that is used to identify patterns of perceptual input. My mental models then enable me to re-cognise my surroundings using past experience. That is, I cognise again, I mentally recreate my surroundings. This is a thoroughly active process, not at all a process of passive reception. But what then? How do we get from re-cognition to action?

2.4.   Models in Use

Since we would normally describe what goes on in a brain as thinking, what I am investigating here is a way of thinking about thinking. The use of the idea of mental models as a way of thinking about thinking was first proposed by Kenneth Craik in his book *The Nature of Explanation* published in 1948. It is now a well-established set of ideas. As I hope to show, it is both a useful and powerful way to develop understanding of thinking and learning.

The significance of Craik's proposal is that not only perception but action too can be based on the notion of an internal model. Imagine a small robot, a three-wheeled tricycle, with a steerable front wheel, able to move around on the surface of a table as illustrated in Figure 2.2.

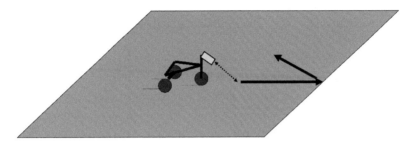

Figure 2.2: Craikian Automaton – mode 1: detection

To prevent the robot from falling off the table, the simplest strategy is to build on the front of the robot a detector which can perceive the edge of the table when the robot reaches it. As the robot moves forward, this device continuously checks for the edge of the table. When the edge is perceived, the front wheel is turned so that the robot moves away from the edge. I suspect that this is the mechanism that most of us would come up with if asked to design such a robot. It is certainly the most obvious way I might think that action is produced and examples of this technique can be found in the natural world. There is a protozoan that swims by means of a flagellum which beats in such a way as to pull the cell along behind it. When the flagellum hits an obstacle, it bends. This triggers a change in its movement so that it moves the cell in a different direction. In a similar way to the robot, the protozoan perceives an obstacle then changes direction to avoid it. (Maturana & Varela 1998)

However, there is an alternative way of designing the robot which makes use of an internal model. This is illustrated in Figure 2.3. The second strategy is to build a more elaborate robot. On the tricycle I construct a model table, and on the model table a very crude model robot. I gear the model robot to move across the model table at a rate proportionate to the rate at which the robot moves across the table, and so the robot 'knows' when it nears the edge of the table and can steer accordingly.

Perhaps it is not an easy task to design mechanical linkages required as I have described, but with today's electronic computing capability this could certainly be done. As long as no one knocks the robot, and the model table remains aligned to the real table, the robot will stay on the table. Of course, in time, errors will accumulate, or someone will knock the robot or the table. Then eventually the table and model table will become misaligned and the robot will fall from the table. But the important point I wish to make is that providing the robot with a model does enable it to take appropriate action on reaching the edge of the table. But not only that, the model could also be used by the robot to predict that action will be required before it actually meets the edge of the table.

Central to the success of this design is the relationship between the model table and the actual table. There are two important factors: the alignment and the shape. Again we see that the *purpose* – to keep the robot on the table – determines the *relationship* between the model and the actual entity, and this is manifest in the attributes that are necessary for the model to fulfil its purpose: alignment and shape.

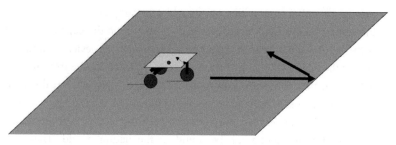

Figure 2.3: Craikian automaton – mode 2: modelling

In comparing these two approaches it can be seen that the information processing requirements of the two are quite different. In the first case the robot requires a large capacity to process information from the outside world and a minimal internal processing capacity. In the second case, it requires a minimal capacity to process information from the outside but does require a sophisticated model-building apparatus internally. If I incorporate into the design of the second approach something from the first approach, then I can get the best of both and overcome the problem of misalignment. The robot can then check occasionally that the real table and model table remain aligned, and if not, realign them.

Close your eyes and move about a familiar room. You won't need your eyes open all the time to move successfully around the room. But you do need occasionally to check your alignment.

Since mental models give rise to action, they can be very important indeed. A friend in his later years was having trouble with swallowing food. He went to consult a doctor. After examination the doctor told him that it was clear that he had a blockage in his oesophagus. Let's call this model A; the action dictated by this model logically follows. My friend was told that he would need an operation to examine and remove the blockage. Not relishing the idea of an operation on his throat, my friend decided to consult a second doctor without mentioning the first consultation. This time the explanation, model B, was different. Because my friend was a writer and had spent many hours over the years hunched over his desk the muscles in his neck which aided swallowing had become weakened and needed exercise. The action dictated by this model is very different from model A, and much preferred by my friend. Exercising his throat muscles alleviated the problem and saved an unnecessary operation.

Models in the medical world are indeed very important. Models of cancer have changed significantly over recent years with a similar consequence of removing the perceived need for drastic operations. (*New Scientist* Feb 2013)

In any complex situation in management, medicine, ecology, economics or politics, understanding the underlying models that people are using in proposing their favoured course of action is of paramount importance. Researching available evidence and testing how well the model fits that evidence, before taking action based on the model, seems a sensible course to take, but in politics particularly, one not often taken.

2.5. Models in Nature

So far my examples have been taken from human experience, but these ideas have a much wider applicability. The human species is just one species with a brain and nervous system. Any animal with a brain and nervous system must equivalently hold patterns and models, and use them to recognise patterns of sensation of light, sound, touch, smell, taste and movement impinging on the animal concerned. More than that, the example suggested by Craik illustrates how an internal model in a machine can be used to guide action. It suggests that having models, as in the second approach, and using the senses to check alignments, as in the first approach, produces at a given moment what is needed to navigate the world. The purpose of the models is simply to achieve survival.

Survival is a necessity for all living creatures. Having internal models and checking alignment reflects the explanation of evolution produced by Darwin. Those with the models best aligned with their external environment have the highest probability of survival. The key is in this relationship between the model and the entity modelled. A simple example is a bacterium that can detect a chemical gradient over its body length so that it can swim up the gradient to where food is plentiful. This can be interpreted as the bacterium containing a model whose purpose is to enable it to find food. A tick sitting on the branch of a tree (Figure 2.4) detects by sensing a particular chemical in the air around it that an animal suitable for feeding is passing beneath the branch. On recognising that signal the tick falls onto the animal. The pattern matching of the chemical to the action of leaping we can regard again as the use of a model to take appropriate survival action in exactly the way I have discussed. The tick exchanges an environment with no food available for one with food available, thus, from the perspective of the tick, managing its environment. Extending these ideas to more complex activities will be discussed later.

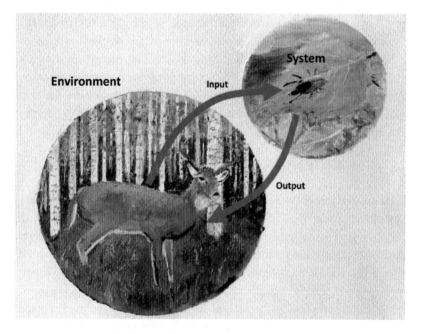

Figure 2.4: *Ixodes scapularis* managing its environment

In more complex organisms, models not only drive behaviour but are used for communication. This is exactly the situation that I described at the start of the chapter. The architect's model is constructed to communicate the form of the building. The building plans are drawn both to communicate and to drive behaviour.

Karl Von Frisch (1967) describes bees at work. The nectar collection system is illustrated in Figure 2.5. When a bee returns to the hive carrying nectar from a newly found source it needs to communicate this to other bees. Bees communicate using a model of where a food source is located. They dance a special dance usually performed on a vertical surface of the hive, communicating to other bees the direction of a potential food source and its distance from the hive. The distance of the food source from the hive is represented by the proportion of time the bee spends wagging its tail in the dance and the direction is represented by the angle to the vertical the bee adopts for the wagging portion of the dance. The spatial location of the source is modelled so that communication of this information can take place. The bee converts its experience into the dance and the watching bees reconvert what they see into action. The internal model is converted to an external manifestation and then reconverted to action by watching bees.

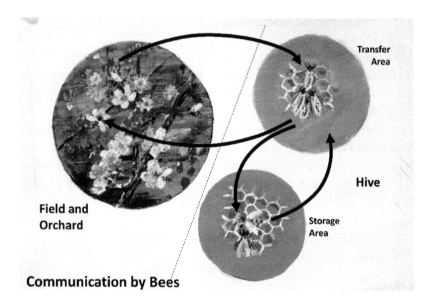

Figure 2.5: Models in action

If a bee arrives back at the hive loaded with nectar and there are no available bees around to unload it and take the nectar back to the storage area, then it does a different trembling dance whose purpose is to attract more helpers to the storage task. These two dances keep the collection system in balance.

If a living system or a machine cannot change its internal structure, the internalised models of that individual system cannot change. In that situation, individuals cannot learn. A nervous system enables an individual animal to change its internal structure all through that individual animal's lifetime, enabling changing internal models. As a result, these individuals can learn; that is, they can change internal models to match a changing environment. Clearly this is a necessary development to enable longer lifetimes for more complex individual animals. In order that a living system can survive if the environment changes, internal models must change. And of course, changes in individuals do happen from generation to generation. Internal structures can and do change through the mechanisms of evolution to enable adaptation to a changing environment.

In the examples I have used so far I have discussed just a single model, but the ideas will need further exploration; a single model is one thing, but the complexity of most animal environments and their brain/nervous systems is quite another.

2.6.   Purpose and Relationship

The *purpose* of a model and its *relationship* to the external entity with which it is associated has been a central theme in the discussion so far. I propose that this is much more important than normally recognised. This can be illustrated by thinking carefully about illustrations written long ago. Approximately 2,500 years ago Plato (in *The Republic*) described a conversation between Socrates and Glaucon. In it, Socrates observes that a bed or a table can come in many shapes 'made by the carpenter' but there is only one 'bed-in-itself' in nature, a perfect form, and in the same way only one perfect form of table. Desmond Lee translates the conversation as follows:-

> 'Then let us take any set you choose. For example, there are
> many particular beds and tables'
>
> 'Yes'
>
> 'But there are only two forms one of bed and one of table'
>
> 'Yes'
>
> 'Then we normally say that the maker of either of these kinds
> of furniture has his eye on the appropriate form when he
> makes the beds and tables we use; and similarly with other
> things. For no craftsman could possibly make the form itself,
> could he?'
>
> p.361

What is interesting here is the translator's word 'form'. In older translations (e.g. Jowett 1894) the word 'idea' is used in place of the word 'form'; 'the idea of a bed'. In a more modern translation (e.g. Robin Waterfield 1993) I find 'type', 'the type of bed', is used. Translating from one language to another is difficult, from a past age to our own even more difficult. The question is what does Plato mean?

The translation goes on:

> 'God then created only one real bed-in-itself in nature, either because he wanted to or because some necessity prevented him from making more than one; at any rate he didn't produce more than one, and more than one could not possibly be produced.'
>
> 'Why'
>
> 'Because, suppose he created two only, another would emerge whose form the other two shared, and it, not the other two, would be the real bed-in-itself.'

The essence of this is that a bed can have many manifestations but these manifestations all perform just one function, that of providing a sleeping place for a person. Thus for me Plato's 'bed-in-itself' is contained in the relationship between a person and something in their environment. That is, some thing(s) in a person's environment can serve as a sleeping place – function as a bed – for that person then and there. So the ideal is defined in terms of a relationship. The same reasoning applies to tables.

I propose that what is meant here by the word 'form' is that particular relationship which I (or any other person) have with an object when I engage with it. I recognise a tree stump in the wood as a table or a pile of hay as a bed just as much as any manufactured table or bed. That essence of bed is contained in the way I relate to the situation, the *'purpose'* that I assign. I perceive a situation as a result of the inputs falling on my senses and react in a particular way, engaging with the situation in a purposeful way by forming a particular relationship. My remembered model of suitable sleeping places is synthesised with the external situation I perceive exactly in the way I have outlined above.

Plato goes on to draw distinctions between, first, a picture of a bridle and bit reproduced by a painter, secondly the bridle and bit as perceived by the makers, but most importantly, and thirdly, the bridle and bit as perceived by the rider; that is in use.

'The painter may paint a picture of bridle and bit.'

'Yes'

But aren't they made by the harness-maker and smith?'

'Yes'

'Then does the painter know what the bridle and bit ought to
be like? Isn't this something that even the makers – the
harness-maker and smith – don't know, but only the
horseman who knows how to use them?

p.367

It is only those who engage with the bridle and bit in action who develop the
essential relationship, so it is only they who truly know what the essence of a
bridle and bit is.

In conclusion I deduce that when I need to lie down to sleep, I first
recognise a suitable sleeping place using my remembered model of the
relationship I must have with a suitable sleeping place. This remembered
relationship is not an abstracted picture of me on a bed, but it is very visceral,
personal and active, enfolding memories of the sensation of lying down,
tucking in and falling asleep. I then again use the model to take the action
necessary to prepare to sleep. The recognised pattern of this suitable place,
which enables the relationship needed, is then given the name 'bed' as a
shorthand label for this relationship when I need to communicate with others.

2.7.   Process and System

The consideration of a mental (or internal) model producing action leads us
to a subtly different conception of what is meant by a mental model. The
considerations that I have described so far lead me to propose that a mental
model is a structure contained in a nervous system or brain that receives
inputs and processes those inputs to produce some output. It is a model in the
sense previously defined. First, it must represent aspects of the external entity
in order for the external entity to be recognised. This representation is
analogous to the representation that enables an 'A' tuning fork to react to the
note 'A' being played on a piano or other musical instrument. These internal
structures enable recognition of the perceived patterns and then, secondly,
potentially transform them into action.

The 'bed' model enables the recognition of the place to lie down to sleep
in order that we can lie down to sleep. Mental models are the result of
accumulated previous experience. So, the development of the models with

which I interpret the world outside is what I mean by learning. I use my experience to build internal mental models, then use these mental models to navigate the world. But my mental models, being internal structures, are not just those developed by me in my lifetime but also those contained in my genetic inheritance. At any moment I react to my environment with combinations of my genetic inheritance and my own personal learning.

Mental models, then, are processes that recognise and transform a particular flow pattern of perceptual input to a particular flow pattern of response output, as illustrated in Figure 2.4. The purpose of the model is to recognise a particular environmental pattern of inputs and to produce appropriate outputs to maintain an appropriate relationship to the environment. It has always to be remembered that this is a dynamic conception, although the words I use don't always convey that meaning. In English, words seem to lose their dynamic meaning; for example a word that is very important here – relationship – is a dynamic process, not at all something static.

2.8. Communication

Figure 2.5 showed an example of communication in bees. Of necessity, to function together, individual bees share models through their genetic inheritance. The returning bee uses a model to produce action to dance a particular pattern. That pattern is perceived and interpreted by the watching bees in the hive and leads to their action. A pattern of signals that an animal needs to interpret and is essential to survival often arises from the broadcast from another similar animal in this way. This use of a model to produce a pattern of action which is then perceived and interpreted by another animal using its models is of course exactly what we mean by communication.

In 1949 Claude Shannon and Warren Weaver published a work on communication which is now considered seminal. Figure 2.6 shows their symbolic representation of communication wherein a communication source selects a desired message and a transmitter translates that message to a signal that can propagate to a receiver. The receiver reverse-translates the signal to a form which can be perceived by the recipient exactly as described above in the example of bee communication.

Here the first bee is arriving at the hive from its scouting trip. The communication source is the remembered model of the whereabouts of a source of nectar and pollen. The arrival at the hive from its scouting trip triggers the bee to action: to become the transmitter and produce the appropriate dance pattern. The channel then is the space between the first bee and a watching recipient bee, the noise the visual interference of other bees going about their other tasks. The watching bee, the receiver, receives the visual signal, and this signal then becomes the input to the model, the destination, which produces the action of the second bee flying off to the source intended.

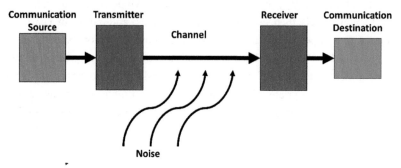

Figure 2.6: The Shannon Weaver Communication Model

Perhaps we would normally think of communication in a human context. Then this most often would entail someone using their larynx and vocal chords to transduce thoughts that they wish to communicate to sound waves which then travel through the air to the recipient's ears, the mechanisms of which transduce the sound waves to their thoughts.

The air in this case is the communication channel, which could involve a telephone link, but then the transmitter is the combined voice plus telephone and the receiver is telephone plus ears. There is almost always noise interference, extraneous noises at each end, and electromagnetic interference in the cables or wireless transmission which will interfere with the ability to hear the intended message.

The discussion in the preceding sections shows that there is an important element missing from Figure 2.6; that is the models used to create the message and those needed to receive the message. There is an assumption contained in Figure 2.6 that the source models and the recipient models align, as they do in my bee example, but in human society there are many occasions of mis-communication because the source model has no exact counterpart in the receiver's model repertoire. Therefore, I need to enhance the model of the process of communication to that represented in Figure 2.7. When the model repertoires do not align, the received message may be very different from the message that was sent. If there is no common language, perhaps nothing like the original message will have got through.

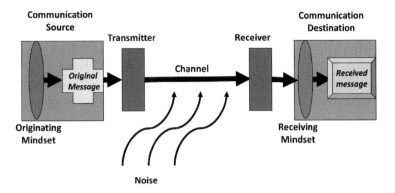

Figure 2.7: The Enhanced Communication Model

If I now come back to Plato's writing I can now understand that whilst he wrote on the basis of his models, those who are translating Plato's communication to us are interpreting Plato's writing using their models. Furthermore, I am reading these translations using my models, and therefore there is no way of getting at exactly what Plato meant because the writing was produced from within a culture that has long disappeared. All that can be done is to interpret the Ancient Greek words through our modern Western receiving mindset.

2.9. Models and Relationships

When I first encounter someone new, I start to build a mental model of that person from my past experience of other people, both good and bad. But then this initial model is modified by the communications I receive through all my senses from that new person over time from the subsequent interactions. They, of course, start to build their mental model of me in the same way, starting from their past experience and then modifying the model of me from the communications they receive from me. I am passing quickly over the word 'interactions', which involve two-way communications, a feedback loop. This is something for my next chapter, and of course is inherent in what is meant by relationship.

When I think about the relationship to this new person it is useful to realise that this is done in the context of my translation of the communications I have received from them, and their translation of the communications they have received from me. This feedback loop stands on little solid ground and can be readily influenced by the way I react to them and communicate with them. The realisation that I can set and change this evolutionary dynamic is a very useful outcome of thinking in terms of mental models.

In this chapter I have briefly discussed the modelling of thinking in terms of mental models, the justification, how the models might work, and some of the implications, using different examples. Central to this discussion are the ideas of purpose, relationship and the pattern that these models follow: that of an input signal or signals being processed to form an output signal or signals. In the next chapter I will look at this pattern in more detail.

3. Understanding system, process and relationship

The expectations which order our appreciated world are rules derived from regularities which we abstract from experience. As a simple example, to sighted creatures whose lifespan is measured in years, one of the most conspicuous regularities of nature is the alternation of day and night. …The abstraction of rules from regularities like [this] is an example of that capacity for pattern recognition on which we rely not only in everything we do but in building the representation of our manifold contexts within which we live. This is the process of discovering order in – or imposing order on – the environment, which science has carried so far.

(Geoffrey Vickers 1972)

3.1.   Constants in the Flow of Experience

As I said in Chapter 1, the writings of Heraclitus of Ephesus are mostly lost, but some fragments have survived which indicate that his philosophy was based on the idea that 'all is flux'. He believed that the world we perceive can only be understood in terms of continuous change, and is most famous for the assertion that you cannot step into the same river twice. This statement, like others of his surviving statements, is a seeming paradox. What does Heraclitus mean by the same river? If I walk down to the nearby river, find a suitable place and step into it, and then if I make the same journey on the following day, the river does indeed look unchanged. The flow pattern on the surface and through the vegetation looks much the same as it did yesterday, so I can assert that this is the same river that I stepped into yesterday. Even if I take no account of the banks of the river or the countryside around there is a constant in the flow pattern of the river; it is the same river. However, if I could identify the water I stepped into yesterday, perhaps by colouring it, the colour would have long vanished, along with anything that was being carried along in or on the water. The water I step into is certainly not the same water. The water in the river continually flows past this spot, and in this sense it is not the same river that I stepped into yesterday.

Figure 3.1: The Flowing River

The flowing river is a useful metaphor for what I perceive with my senses. I experience the world as a flow of sensory experience like the flow of the water in the river. My eyes, ears, and nose report ever-changing patterns of light, sound and smell like the water flowing past. We each experience our world outside as this flow pattern of sensory experience that we have to make sense of in order to survive. It is the identification of invariant patterns in the flow of water that indicate that it is the same river, just as it is the invariant patterns in the flow of sensory experience that indicate that it is the same world. To build an understanding of that world we must look for and identify invariant patterns within the flow of our sensory experience to appreciate the same bed, the same table, or the same person.

Perhaps you can imagine a newly born child gradually recognising the sound, smell, touch and sight patterns that are its mother and integrating those with the already familiar sound of a voice (Sinha 2013). A mother is one of the very first constants of a child's life. In a completely different context perhaps you can also imagine the flow of materials, communications and people marking out a manufacturing organisation just as in a film taken with time lapse photography. In a manufacturing organisation, raw materials flow in, finished products flow out, there are flows of communication. The shape and form of the buildings are like the rocks on the riverbed causing particular flow patterns. This all can be seen as a particular invariant but dynamic pattern that is the organisation.

In the last chapter I discussed what Plato meant by 'a bed in itself' and the word 'form' as translated from the original Greek. What is important in this discussion is that it is my relationship to my bed that is a constant in my stream of experience. Every evening when I go to bed I reproduce at least in general terms the same experience. I perceive the same things and I make the same movements. As I have discussed, the perceptual inputs and movement outputs are what I mean by my relationship to my bed. But, just like Plato, I have seen many other beds and am able to recognise a bed in many different forms through this constancy of relationship. Recognising this further, more abstract, constancy from all these experiences of different beds, I have been able to appreciate the idea of a bed, as discussed by Plato. From my experience I first abstract the idea of my bed. From the experience of many beds I make a further abstraction of a bed, an idealised constant pattern from my wider experience. The relationships will not be exactly the same but there is a commonality of purpose to all these relationship situations.

This wider experience of many beds enables me to abstract what I regard as the ideal bed from the similarity of relationship; and the same applies to the idea of a table, or any other object to which I relate. Heraclitus chose a flame as his archetype (Popper 1963), something seen as a thing but which is clearly a process. But further abstractions can be made; in particular there is a common pattern across all relationships. At the end of the last chapter I proposed that the common invariant pattern across any and all mental models is an invariant pattern in a flow of purposeful input→process→output; in this way I identified the pattern of system as common to all. It is this pattern common to all experience that I wish to explore in this chapter.

3.2.   Purpose and System

If then, these models are best understood as systems – that is input→process→output structures – the questions I would like to pursue now are: what are the essential features of a system and what are the modelling implications? After all, if this is the way I understand that a nervous system functions, as an input→process→output structure, wouldn't it be a good idea to explore that same idea of using a system as a way of building understanding of the world? Is this pattern useful and if so, where does that lead? This exploration seems to me to be exciting new ground.

The word 'system' is in common use but can mean many different things. The most commonly used definition of a system goes back a long way (Bertallanfy 1968). It says that a system consists of parts with dynamic interrelationships and interactions between those parts. But thinking about only what is internal to a system is missing what to me is of primary importance for any system: its relationship to the external environment. If I view a system as an invariant pattern in a background dynamic flow, the flow and the invariant pattern are both essential to my thinking. It is this, a system (input→process→output), that I discriminate from the background flux by conceptually marking a static boundary. But most importantly the internal world of the system, that which lies inside the boundary, is fundamentally connected to the external world outside the boundary through the input and output flows. Even though I might focus on an invariant pattern in a dynamic flow, the flow is there and essential to the analysis. At this point in the exploration I am not concerned with possible parts or their interrelationships; that will come later.

The point of primary importance now is that the flows are effecting changes in the world outside the system boundary. At some future time the world outside will have added to it whatever the system is producing as its outputs, and subtracted from it whatever the system is requiring as its inputs.

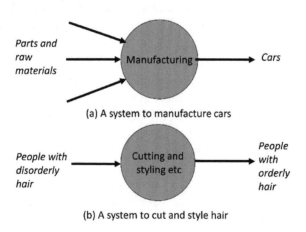

(a) A system to manufacture cars

(b) A system to cut and style hair

Figure 3.2: Examples of a system

Therefore, in the way I have conceived it, a system is purposeful; that is, I ascribe to it the goal-seeking nature of producing an output from the process of transforming the input. So, for example, in my observing of the world around I see that a plant manufacturing cars is evidently a system whose purpose is to produce cars, and a farm is a system whose purpose is to produce food. The system's purpose is the starting point to begin to think about any particular system, but it is worth noting that in our usual way of thinking, clear purposes are not necessarily self-evident. For example, what is the purpose of a school, a bank, a wedding, a restaurant, the health service, a government (all of which may be viewed as systems in this way)? Whether it is immediately clear or not, in order to define a system model, we need to define the system purpose at the outset in order to construct the system model. This is a fundamental point which early systems theorists were slow to realise because it is not our usual way of thinking.

Thinking about systems in this way changes the way I think about the world and enables more precision in thinking. I certainly grew up believing that the tiger in the zoo was the same animal as the tiger in its natural habitat. The focus of this deduction is on the shape, colour and form of the tiger, cat shaped, stripey, four legs, etc; that is, I conceived the tiger as an object in the same way that I think of a bed or a table as an object defined by attributes – two metres long, four legs, etc. But the tiger of each of these cases is not the same system. If I focus on the relationship between the tiger and its environment it can be seen that there are great differences between the two situations. In the first case the purpose of the tiger is to be a supreme attraction for people to the zoo, in the second to enact the purpose of supreme predator in the wild. The relationship between the tiger system and its environment – the way in which the tiger system interacts with its environment – is very different in these two cases. Modern zoos do try to reproduce as far as possible conditions similar to the wild, because all animals have their in-built species models as described by the quotation at the head of Chapter 2. Like any animal, a tiger has expectations of its world and ways of relating developed over thousands of years of evolution which a zoo cannot change. However, the caged tiger is not the free agent that it would be in the wild depending on its own resources to survive. That essential relationship of tiger to environment is fundamentally different in the zoo from what it would be in the wild, and therefore the two tiger-systems are different. What I highlight here is the essential difference between thinking of a tiger as an object defined by properties of the object itself and thinking of it as a system that is defined by its relationship to the environment in which it exists.

This is an essential difference that will be discussed in more detail later. Suffice to say that the idealised way of analysing and describing our world as objects became highly developed with 18th- and 19th-century physics but the biological world is a world of systems. The modern biological approach is that all animals are in fact ecosystems, a complex mix of bacteria and animal (Arnold 2013). In the biological world there is a dynamic relationship between any individual and its environment and there must be what we might call a *productive relationship* between them for it to continue to exist. Both the environment and the individual must gain from the individual's existence.

Whilst defining purpose does give me increased precision in my thinking and begins to point towards an understanding of relationship, it can lead me astray. I am indebted to a colleague for pointing out that the nuclear power industry likes to style itself as 'a system to produce electricity without adding carbon dioxide to the atmosphere'. But the nuclear power industry at present can equally be styled 'a system to produce waste toxic to life which cannot be safely disposed of'. Both are outputs of the same system. 'Purpose' is a vital tool in defining a system but in terms of the relationship of a system to its environment this example is an illustration of the need to be careful. A singular defined purpose may not be the whole story. A purpose at most defines output of the system, and in this example only part of that. In considering the relationship to the environment, I must consider the totality of output and the totality of input, or perhaps, as much as is humanly possible.

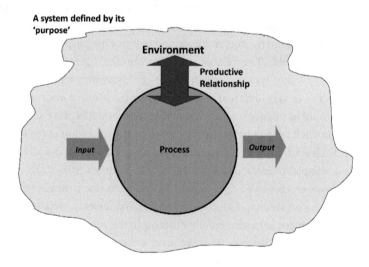

Figure 3.3: System Features

Humberto Maturana (Maturana & Varela 1980), a biologist, introduced the term *structural coupling*. He viewed a system as structurally coupled to its environment through its input and output. Whilst this term has the advantage of including all input and output, it has the disadvantage of easily forgetting the purposeful nature of a system. It must not be forgotten that the system's purpose is included. Neither the approach through defining a purpose nor the approach through structural coupling gives the full extent of what is required in defining a system relationship. Neither approach is complete, but they are complementary and an aid to one another. Fundamentally what characterises a system is its effect on its environment which is both dynamic and teleonomic (that is, it has a direction).

3.3.   Changing relationships: perspective, boundary and structure

It is always possible to explore any complex situation in terms of a particular system attribution. That is, I can knowingly take different perspectives on the same situation in this way. For example – what is the purpose of the health system? Most would answer that it is to make well people who are sick or broken in some way: a repairing system. But many think that it ought to be also a preventative system. These different purposes give rise to different requirements. A preventative system would need to focus on education of the general population. It is much more about individuals in the general population taking their own actions on the basis of their own understanding, whereas a repairing system is a system where things are done to individuals in the general population by knowledgeable specialists.

A privately owned health system is different again because of the onus to put the purpose of profit as primary, a requirement which necessarily focuses on sales – perhaps of drugs, perhaps of questionable or unnecessary procedures – and necessarily takes resources out of the system as dividends. The relationship between the environment and the health organisation itself is different in each of these cases. They are not the same system even though we label each a health system and often they are talked and written about as if they were the same. It has always struck me that this is the source of much political argument, because unknowingly in discussion people have taken different perspectives and each is basing their argument on a particular conception of the system they are discussing. They are not the same but actually different systems, and therefore it is never possible to reach agreement. Being aware of the differences in purpose, and therefore requirement, is essential for constructive discussion.

Different perspectives are everywhere in our thinking about our world. I could characterise our car manufacturer in Figure 3.2 as 'a system to make quality cars for customers' or equally 'a system to provide a good return on capital to shareholders' or 'a system to provide skilled fulfilling employment' or even 'a system to maximise the take-home pay of senior managers'. Again, these are similar but different systems. If I were to analyse each in a logical way, I would find different emphases within the system on the various activities required to achieve the purpose designated. That these differences can be vital was graphically illustrated by the 'Railtrack' company, which was responsible for the UK's network of railway tracks. For a few years it was an effective system to provide a good return on capital for investors, but then failed completely because it did not sufficiently resource track maintenance.

The process of track maintenance is logically a primary part of a system to provide train services to customers, but logically secondary in a system to provide shareholder profit; perhaps something to be aware of in the context of health systems.

The British railway system also provides an interesting illustration of the way in which purpose and boundary are connected. On moving the system from public to private ownership the change in the purpose of the system, moving away from providing travel services for customers and fulfilling employment, towards providing returns for shareholders and larger remuneration packages for directors and senior staff, can be seen in the experience of travellers. It is characterised by a change in the position of the conceptual boundary around the system. On privatisation, many rural stations were closed or ceased to be maintained. Activities that were within the system boundary moved outside, either ceasing to exist or ceasing to be the responsibility of those managing the system. The indication is that a change in the environmental relationship necessitates a change in boundary and internal structure and also, of course, a change in boundary and internal structure will change the environmental relationship. The privatised railway system has a different relationship to its customers from that which pertained previously.

In Chapter 2 I explored the way in which internal models enable perception and action. What an animal perceives and how it reacts to those perceptions is the basis for the particular relationship that the animal has with its environment. If the animal changes its relation to the environment then I would expect a change in structure. Sure enough, animals do change their relationship to an environment, and their structure in all sorts of ways; insect life-cycles are a perfect example. Most people will be familiar with the caterpillar-butterfly. The necessities of its life-cycle require a complete change of form. The resulting metamorphosis from caterpillar to butterfly results in structural changes and a complete appearance change that takes place within the chrysalis as the caterpillar mutates into a butterfly. This metamorphosis enables the increased mobility required for an increased variety of opportunities for reproduction.

Such a drastic change of physical structure is not necessary in more complex animals with larger brain/nervous systems. A more complex brain/nervous system can provide a variety of different responses. A change of model in such an animal can be provided within the complexity of its brain/nervous system. This seems to me to be exactly the purpose of the brain/nervous system. It seems to me no coincidence that the animal with the greatest adaptability, *Homo sapiens*, has evolved the most complex brain/nervous system. A person moving to a new environment has the opportunity to learn and build new relationships, as a result mankind can be found in almost every part of the planet. How this can be modelled will be the subject of following chapters.

My conclusion is that whatever the system, its relationships to its outside environment, which are characterised by its purpose, determine the internal structure of the system and vice versa. Therefore, for a human organisational system it really matters that those who work within it have a clear understanding of its purpose. The differences here are much more than the difference between the two tiger systems. At root, an organisation, for example the health service, is a collection of people and the relationships they have to each other, various external stakeholders, and sources of information internal and external. What I mean by relationship here as elsewhere is all the exchanges of input and output that take place. I can imagine some sort of picture as illustrated in Figure 3.4 which depicts a set of people relating together. The people within the boundary are those within the organisation.

Suppose I am appointed Chief Executive Officer and reorganise to make the organisation more 'efficient'. In my experience this usually would mean that the primary focus is on profit or costs. Without awareness of the way in which structure and purpose are intimately connected, unknowingly in my zeal I change the purpose of the organisation, and the boundary moves to include or exclude processes in or from the system. The changes I make in the internal structure result in the purpose moving away from any customer focus. People complain vociferously at the ineffectiveness, I am replaced and a new CEO is appointed, and so the cycle continues!

Understanding that structure and purpose are intimately linked is essential for any manager. I can also envisage that the purpose of the whole system is an emergent property of the purposes of those acting within the system. But this is not a one-way relationship: the influence of those acting in the environment constrains the purposes of the actors within the system, so for any sizeable organisation the complexity is beyond the understanding of a single person.

Purpose and boundary

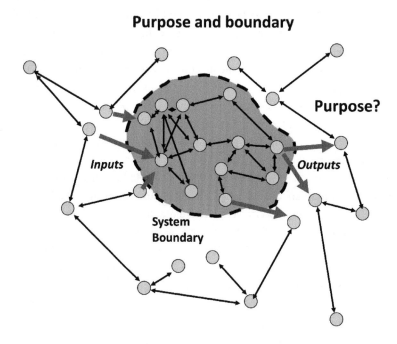

Figure 3.4: An organisation of interacting individuals

In summary, the final point in thinking about purpose is that it is a way of characterising a system and to a great extent defines the relationship between system and environment. If a particular system continues to exist, then it follows that the purpose must be one which has some advantage to the environment; in some sense, it must be a *productive relationship*, both to the system and to the environment. This point is one which is vital when thinking about systems in the natural world where 'sustainability' would be the word used to describe this mutual support, which couples system to system environment.

This concept is also important in the consideration of the purpose of government. Whatever the collective of people being governed, sustainability of the government is, I would propose, a first consideration. A government must maintain positive external relationships to avoid interference from outside, and also maintain positive internal relationships to avoid internal breakdown. I observe that the maintenance of both these kinds of relationships is, at the time of writing, becoming more difficult because the relationship between people and the natural environment has not been maintained adequately. This is a subject I will return to in Chapter 8.

3.4.   States and Variety

Any musical instrument is an example of a simple system. We may very well disagree on the music we like but I hope we can agree that the purpose of a musical instrument is to produce a sequence of sounds which give people pleasure. When I play my piano, the input to the musical instrument is the energy from my hand movements in producing vibrating strings (or in the case of other instruments it could be air blown in from lungs or bellows, or even combinations of these various techniques). There are many ways of achieving the input but the output is the patterns of sound I require. I will fix for the moment on a violin as my example with its four sound-producing strings, but the underlying principles that follow apply to all instruments. A string vibrates in a particular way with what physicists call a set of modes which depend upon various properties of the string (examples of which are illustrated in Figure 3.5). Each mode of vibration, or state, produces a particular musical note, but in use often the string vibrates simultaneously in various combinations of these modes, so called overtones. Each of these possibilities is a possible state of the string system. What is important here is that the number of possible states defines the *variety* of the system. Even for the violin, which is, on the face of it, a simple system, the number of possible states of the system is large. A violin has four strings each of which can vibrate in a multiplicity of combinations of modes. Then add in the violinist.

Now I must consider the change in the length of string produced by the violinist's fingering hand, changes in pressure applied to the strings by the bow and the fingering hand. This then gives some idea that the variety is very large indeed. The violinist chooses particular configurations of fingering and bowing to manage the sound output; that is they reduce the *potential* variety of the multiplicity of states to the relatively few that they actually use to achieve the output required.

For a second and much more complex example, consider a class of school children. Because this is a rather complex situation, fixing precisely the system under consideration is not as straightforward. For the moment, I will look from the outside and define the purpose of this system as: to increase children's knowledge, skill and understanding during the course of a school day. At first this might seem a reasonably simple situation, familiar to most people from their past experience. A class consists of perhaps thirty students with different backgrounds, home situations and understandings. At any one time in the classroom there will be a very large variety of possible activities, visible and invisible to an observer or the teacher; for each of the thirty students – daydreaming in many possible ways, talking to a classmate about many possible things of mutual interest, getting on with work in many possible ways, even disrupting the class in many possible ways. The teacher in managing the class has the task of reducing this enormous *variety* of potential activities, the variety inherent in the situation, to those activities that are conducive to the learning that s/he wishes to take place. This is the problem of *managing* the class. It is an example of the problem of managing, just like managing the sound output for a violin, or for managing in any other complex situation.

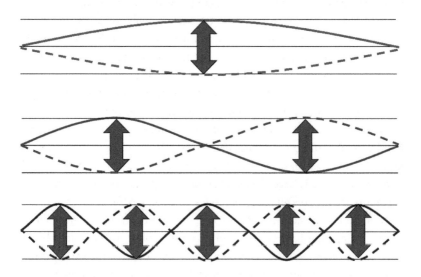

Figure 3.5: Examples of system states for a vibrating string

To manage is to reduce the variety of activities to the set of activities required to achieve the purpose of the task in hand. The traditional approach in the classroom to reducing variety is to isolate children from each other by requiring silence and arranging desks in such a way as to make communication between the children difficult. This does reduce the variety of externally visible unwanted activities, but it does not reduce the invisible activities, nor does it necessarily help in focusing students on the learning activities required. It does reduce for a teacher the *perceived variety*. As I concluded in Chapter 2, without a model there is no perception; perception of variety implies noticing differences, it follows therefore that the variety that any observer perceives depends upon their perception and understanding. I cannot overemphasise this point, it is a most important one. I will come back to a model of learning in Chapter 6 which points to much better ways of managing a class of students to achieve learning.

An excellent way for anyone to illustrate the notion of *variety* is to keep a diary for a day then to reflect on the variety of activities that have been undertaken and to examine the strategies by which variety has been reduced – for example by leaving letters and emails unopened or telephones unanswered.

Each internal state of a system produces a slightly different process of transformation from input to output, and therefore a minor change in relationship, whilst the purpose remains the same. For the violin, the sound output changes as the state of the strings change. For the teacher, different approaches and teaching techniques change the state of the class. Whilst state changes correspond to variations in the relationship of the system to its environment, if the state changes are close to each other there is no overall change of purpose. The concept of variety as the number of potential states of a system is central to the development of understanding the principles of managing a system. Here I use the word 'managing' to stand for any of the activities like controlling, regulating and steering that entail limiting the processes of a system to a particular desired purpose. In this respect the idea of purpose as a directional arrow is useful. In managing a system I seek to achieve a particular direction for my metaphoric arrow, but in many situations the precise state does not matter. However, the direction might very well vary from second to second in maintaining the productive relationship with the environment. An exploration of these ideas is the purpose of the next chapter.

3.5.   Coupled Systems: Feedback

At the end of Chapter 2 I touched briefly upon an example of relationship-building between two people through their communications. The essential nature of this example is that it is an example of two coupled systems. In this case there are two people considered as systems (input→process→output) coupled by the communications between them. Understanding the behaviour of systems coupled in this way is vital in building models of experiences in terms of systems. It is important because almost any system can be analysed into coupled sub-systems.

However before thinking more about people and their interactions I want to discuss a much simpler example. One of the most important advances in the application of steam power was the application of a governor to regulate the speed of operation of a steam engine, exactly an example of limiting the variety of a system discussed in the last section. It is also a simple example of two closely coupled systems each directly affecting the other in a closed-loop feedback relationship.

Steam engine

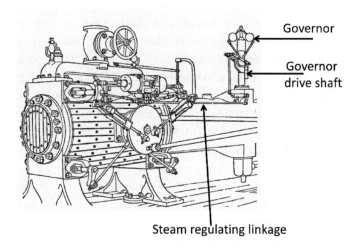

Figure 3.6: A steam engine and governor

In a steam engine, fuel is burnt to produce heat which is used to heat water to produce steam in an enclosed space. The pressure generated by expanding steam is then used to drive machinery. Coal was burnt to produce the heat but the problem was that this process is difficult to control, too much heat and the speed of the machine increases, too little and it slows down. The amount of steam reaching the engine needs much finer regulation than can be achieved by trying to regulate the heat applied. Regulation of the speed was achieved by introducing a governor consisting of two heavy balls attached to a rotating vertical shaft (as shown in Figure 3.6) so that they would fly out away from the shaft around which they were rotating as a result of centrifugal force as the speed of the engine increased. This movement outwards was then connected to the steam supply so that the outward movement would shut off the steam supply hence slowing the engine. As the engine slowed, the balls would fall back towards the rotating shaft opening the steam valve again and enabling the engine to speed up. Correctly designed, the governor-steam engine system would maintain a constant speed even with a variable load.

Connected Systems with Negative Feedback

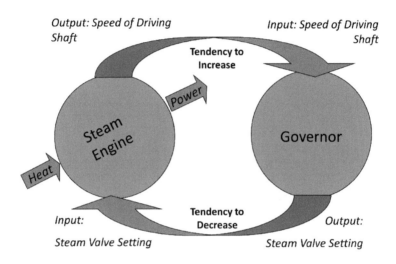

Figure 3.7: Coupled systems with negative feedback

The two systems are defined as having *negative* feedback between them, forming a *negative feedback loop*, because the tendency to increase in one of the coupling flows causes a decrease in the reverse coupling flow. I have illustrated this in Figure 3.7. Even so, the design is critical: any delay in the operation of the feedback would cause the engine to speed up and slow down in an oscillatory fashion rather than operate at constant speed – a condition known as hunting.

Two systems can be coupled in such a way that both coupling flows act in the same direction. The coupling feedback is then said to be *positive*, forming a *positive feedback loop*; the tendency to increase in one of the coupling flows leads to a tendency to increase in the other coupling flow. Alternatively a tendency to decrease in one of the coupling flows leads to a tendency to decrease in the other coupling flow.

A useful example which can exhibit any of these behaviours is two people locked together in confrontation. At any point in the confrontation a proponent can choose to respond to an aggressive approach with aggression, amplifying the level of the confrontation – a positive feedback. Alternatively they could respond with conciliation, reducing the level of confrontation – a negative feedback. A situation of continuing positive feedback will spiral out of control but responding to continuing aggression with conciliation has a chance of stabilising the situation. If the original proponent responds to conciliation with conciliation then the positive feedback runs in the reverse direction reducing and dissipating the confrontation, and possibly promoting friendship.

The Homeostat

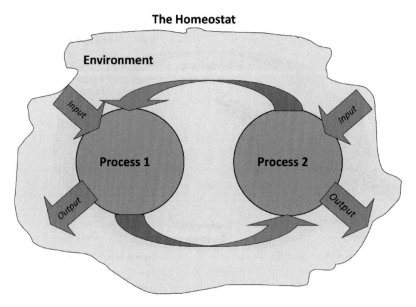

Figure 3.8: Two coupled systems

Many situations can be modelled as two interacting systems in this way. Consider a teacher and pupil. The teacher models the pupil's state of under-standing from the inputs received from the pupil and uses this model to determine the outputs that s/he communicates to the pupil. The pupil uses these inputs to build their understanding, their model, of the subject of the teaching. The pupil responds to questions and challenges with what they think the teacher requires. These responses modify the teacher's model of the pupil's under-standing and therefore modify the teacher's output. Each adapts to the other.

There will be some situations of this nature in which the feedback is continuous and unbroken as with the governor and steam engine, and some where the feedback is intermittent as with the confrontation and the teaching. Figure 3.8 illustrates the general structure. The simplest situation, the archetype of two coupled systems, is a system in its environment, where Process 1 is the system which is the focus of attention, the *system in focus*, and Process 2 is its environment, and each influences the other.

3.6.   Fractal Structure

The most powerful consequence of thinking in terms of a system as a purposeful input→process→output pattern is that the internal structure of any system under examination can be seen as an interacting set of systems each of that same pattern. Any system can be seen as being made up of sub-systems (or indeed as a sub-system of a super-system). The picture that emerges from this analysis in terms of processes is one of a fractal structure.

Figure 3.9: A fractal structure: the Romanesque cauliflower

Fractal structures are self-similar at different scales in exactly this way (Mandelbrot 1982). Many examples can be found in nature, the shapes of mountains, coastlines, the surface of the sea, are all examples of fractal structures. The Romanesque cauliflower (Figure 3.9), ferns and the branching of trees are all excellent examples which show self-similarity at different scales. The importance of this concept here is that, if analysis is conducted using the same underlying system pattern – that is systems are analysed in terms of systems – the final result is a self-similar pattern tracking the pattern of flows on ever smaller scales as if turning up the power of a metaphorical microscope.

I will take a restaurant as an example of this analysis approach As discussed in section 3.3, the first task in the analysis is to fix the perspective from which I will view the restaurant. To do this I must decide the purpose of the restaurant. Since I believe that the goal is to achieve sustainability there is just one option, I define it as 'a system to give the customer a feeling of well-being through eating a meal'. This establishes a productive relationship with the environment. It describes a purposeful input→process→output structure processing customers, whatever the initial state in which they arrive, to a state of well-being when they leave. This changes the environment in which the restaurant sits so customers are likely to return in the future. Also, because of their recommendation, their friends, relatives, and acquaintances will consider coming.

Having set a purpose, I should analyse the internal functioning of the restaurant in the same way, in terms of that same pattern (purposeful input→process→output).

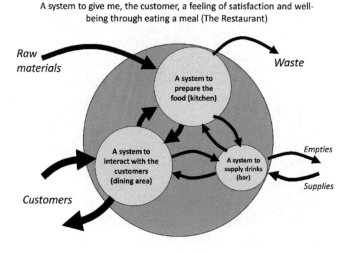

A system to give me, the customer, a feeling of satisfaction and well-being through eating a meal (The Restaurant)

Raw materials

Waste

A system to prepare the food (kitchen)

A system to interact with the customers (dining area)

A system to supply drinks (bar)

Empties

Supplies

Customers

Figure 3.10: A restaurant system illustrating the fractal structure

I view the restaurant as consisting of three sub-systems: a system to prepare the food, a system to interact with the customers, and a system to supply drinks, as illustrated in Figure 3.10. I could now further analyse these sub-systems into their sub-systems, but I will leave that to you for the moment. Hence in this approach to analysis, systems always contain systems themselves. The input flows to the system in focus will be input flows to various sub-systems and again the output flows will be output flows from various of the sub-systems. Interconnecting flows between sub-systems cannot be forgotten since they define exactly the relationships that exist between sub-systems. What I haven't mentioned here is the management of the restaurant. Management will be the subject of the next chapters.

Looking carefully at this diagram (Figure 3.10) you will see that the pattern input→process→output defined by a statement of purpose is exactly the pattern that is used to depict the system of interest – the restaurant itself, and the systems which make up the restaurant as a whole – the kitchen, the dining area and the bar. The three sub-systems have exactly this same pattern of structure as the original system of interest. It is this pattern of structure – one which can be observed at one level and repeated again at another level – that the word 'fractal' or 'recursion' describes. Reaching this point it seems very clear that fractal structures should occur in nature because nature itself is fundamentally processual. Using any modelling pattern other than the one I am exploring here in modelling natural phenomena is neither as effective nor as efficient.

3.7.   Super-system and Schema-system

Hidden in the discussion so far in this chapter are two important ideas relating to the analysis of a system. They are closely related but need to be differentiated. A system can contribute to a larger scale system in two different ways.

The first of these is the simpler. A system is a constituent part of a larger system, which I will refer to as the *super-system*, and interacts with other constituent systems in contributing to the super-system of which they are part. In the previous example the kitchen is a constituent part of the restaurant and interacts with the dining area and the bar in contributing to the restaurant. In a similar way, the brain/nervous system is very different from the heart/lung system of an animal but these two systems interact together and together with other systems form the super-system of the animal itself. In both these examples there are systems with different purposes interacting together to contribute in different ways to a super-system whose purpose emerges from the interacting purposes of the constituent systems.

The second way a system can contribute to a larger-scale system derives from a commonality of purpose explored earlier in this chapter. I proposed that commonality of purpose in the relationship to me across many different individual beds forms the basis of my idea of a bed. I wish to explore this notion of a generic relationship (or structural coupling) stripped of individual details. Similarly, in the system models I construct when exploring the relationships of systems I observe in my environment, I recognise commonalities of relationship through my attribution of purpose. For example, every tiger in the wild has a similar purpose in relation to its environment: a similar structural coupling. This forms the basis of my idea of a tiger in the wild and is in the same way a generic system which is stripped of individual details. This generic model I will always refer to as the *schema-system* to the individual system. A schema is defined as 'an underlying organisational pattern or structure; conceptual framework'. The idea of a tiger in the wild is schema-systemic to any individual tiger in the wild.

Since the internal structure of a system determines its relationship to its environment and vice versa, I deduce that in any analysis of the form and structure of tigers in the wild I will find that they are similar, and of course they are. The same will be true for man-made systems, a collection of car manufacturing plants share a purpose and commonalities of structure. This is not to say they are identical; tigers are not identical and neither are car manufacturers, but there will be commonalities. This will also be the case for any other collection of individual systems where there is a commonality of relationship or purpose. In the case of any example in the natural world this is exactly how we define the idea of a species.

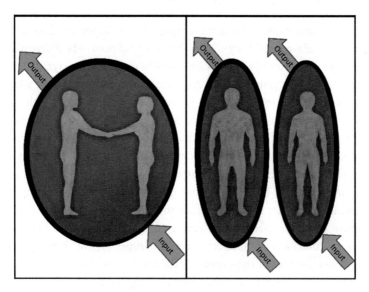

Figure 3.11: Super-system and schema-system

Therefore a schema-system contains the commonalities of purpose and structure across individual systems. It is a common systemic framework that can be extracted from the comparison of two individual systems. It should be noted that there is no necessity for the individual systems to interact with each other. The individual beds of my experience can be far apart in space and time. I can also review my experiences further and begin to categorise them in terms of their commonalities. I can consider some beds *comfortable* and others not. When Plato writes of *beauty*, for me that means there is a quality in the relationship he has with some *thing* in the world, in the perceived purpose and structure that there is in common with other *'beautiful' things*, even though they may be as diverse as a bed and a bridle.

In Figure 3.11, what I am concerned with in the first panel is that there are two different systems interacting with each other which form a super-system: a couple or a family unit. In the second panel I observe the commonality of structure and relationship to the environment that the two systems exhibit. If I compare the structure and form of a man and a woman, I can see the schema-systemic commonality that indicates the species *Homo sapiens*. For me, super-system carries the essence of relationship, while schema-system carries the essence of structure.

3.8.   In Summary

The purpose of this chapter has been to try and define what a system is, and how it can be characterised: the *being* of a system. The being of a system is best thought of as the purpose of the system despite the drawbacks in thinking of it in this way which I have discussed. There is no other way in the English language to ensure that the purposeful nature is immediately obvious, and that at least some of the environmental connection is also immediately stated.

Many authors appear to be thinking in terms of systems in their writing but it wasn't until the 1970s that this approach became explicit. Both Stafford Beer (1972, 1979, 1985) and later Peter Checkland (1981) realised that the extant conventional approaches to Systems Thinking based in natural science and engineering were inadequate in modelling human organisations. Both based their ideas on the concept of a system as a purposeful 'input→process→output' pattern. This resulted in two organisational systems approaches, the Viable System Model (VSM) and Soft Systems Methodology (SSM) which are now widely used in management, organisational and information technological systems analysis. However, I hope that I am beginning to show that this approach has much wider applicability than just in the domain of management.

A system is a model that I create to describe something that I observe in its relation to me. Exactly how I model a system depends on my perspective – as I discussed in Chapter 2 – and this depends on my experiential history If I identify a system in the natural world, from my perspective its being is inherent in its structures as shaped by the evolutionary processes it has experienced in its interactions within its ecological niche. The systemic purpose of a natural system is its sustainability within its particular ecological niche. What is primary to the relationship – system to environment – is just as expressed by Richard Dawkins (1989) for a gene:

> ...the phenotypic effect of a gene is a concept that has meaning
> only if the context of environmental influences is specified,
> environment being understood to include all other genes in the
> genome. A gene 'for' A in environment X may well be a gene
> for B in environment Y. It is simply meaningless to speak of an
> absolute context free phenotypic effect of a given gene.

I might equally write that the resultant effect of using a screwdriver depends on the context, the environment in which I use it – as a lever, a chisel, or as a screwdriver. In both cases, gene or screwdriver, it is the relationship of system to environment that produces the ultimate effect.

For a man-made system like the restaurant, just as for any animal or plant, the schema-purpose is sustainability: the way that is achieved can be decided. Beer in his VSM named the repository of the 'being' of an organisational system System 5, with the caution that 'a system is what a system does' from the *observer's perspective*. It is the *observer* who defines everything. In Stafford Beer's words (1979 p.36) it is "the OBSERVER, who defines the system, and its purposes, and (it turns out) its variety."

This chapter is entitled 'System, Process, and Relationship' but the pattern that I have defined as 'system' includes both the process and the relationship to an environment. Therefore, whilst I read of Process Philosophy and Process Analysis I would argue that these are really misnomers since the essential part of any process is its relationship to an environment. Focusing on the process itself, without including the relationship, misses that essential and fundamental defining part. So from here on I will use the word 'system' and assume that it is understood as the purposeful pattern of input→process→output in an environment, a relationship to the environment defined by the input- output combination.

4. Holding Steady: Structure

> Here, then, is a striking phenomenon. Organisms, composed of material which is characterised by the utmost inconstancy and unsteadiness, have somehow learned the methods of maintaining constancy and keeping steady in the presence of conditions which might reasonably be expected to prove profoundly disturbing.
>
> (Walter B. Cannon 1939)

> Wisdom is to know the thought by which all things are steered through all things.
>
> Heraclitus – as quoted by Diogenes Laertius (Kahn 1979)

4.1.   Introduction

"There is nothing so constant as change" is a saying attributed by Plato to Heraclitus. It seems to me to be quite natural to take the view that all is flux and change, although this is not a common view in the society I live in, as I explored in Chapter 1. Evolution in the natural world is a prime example of flux and change, exemplified by the appearance of new diseases and of new resistances to agricultural pests; and by birth, growth and death in the natural and social worlds. This view of the world around us is not the view of the mainstream of Western thinking. After the ancient Greek philosophers it wasn't until the 1878 publication of Claude Bernard's *Les Phénomènes de la Vie* that it was noted that it was remarkable that in a world of constant change the internal environment of animals remained constant. Walter Cannon in his 1932 book *The Wisdom of the Body* coined the word 'homeostasis' for this phenomenon. But the question that naturally arises from taking the view that all is flux and change is – how is stability achieved? This chapter begins an exploration of how stability, homeostasis, is achieved, using and developing the modelling techniques I have developed in the previous chapters. Three chapters are devoted to this exploration, each focused on a different aspect of the internal structures and processes that are necessary within a system to maintain homeostasis in a changing environment.

The answer lies within the science of *cybernetics*, defined by Norbert Wiener (1948), one of the founders of cybernetics, as "the science of control in the animal and the machine". The word 'cybernetics' has more recently become associated with robots and science fiction, but Wiener originally took it from the ancient Greek for the helmsman or 'steersman' of a ship as in the original Greek words of Heraclitus quoted above. The essential requirement of *steering* a ship is to hold a course out of the many possibilities. On a ship in ancient times this would mean taking account all the dynamics of tidal currents, the wind and the destination sought, and in stormy times the availability of shelter. In the quotation above Heraclitus seems to be generalising this ability to carry out the complex task of steering a ship as a more general metaphor for a definition of a wise person, perhaps someone who can steer their way through any complex situation. Is this a sensible view? Whatever your view, the science of cybernetics provides the general understanding of these situations and this leads towards an understanding of homeostasis.

In the English language, a number of words overlapping in meaning all contain in essence the idea of *steering* – managing, controlling, regulating, governing – and even others like directing, commanding, curbing, restraining. I am here not concerned with the subtleties of differentiating the precise meaning of these words but want to pursue what is entailed in achieving stability in a changing environment and what is needed to do so. If I look around, the people in my world seem the same as they did yesterday, but of course they are not. Body cells have been lost and renewed; they have forgotten some things and learnt new things. Change is only noticed when it is drastic and obvious; but despite all the disturbances, all the dynamics, stasis is achieved. The person I see is an apparently invariant pattern in the flowing river of life.

In general terms, in order to achieve constancy, some form of 'steering' is necessary. What I will write in the systemic language developed in the previous chapter is that what is required is a 'steering' mechanism, the effect of which is to hold a system to a particular purpose, a particular environmental relationship, despite the many and varied environmental disturbances which impinge on that system. The complete story and a full understanding of how homeostasis happens can be elaborated from this starting point in several steps.

4.2.   Variety: Problems of Steering in a Complex Environment

We all have experience of the process of managing: we manage ourselves, our family members, our job and our colleagues in one way or another. We try to hold our course, to maintain relationships despite the many and varied changes around us. In order to explore the difficulties in holding steady in a dynamic environment, I will start by considering the difficulties inherent in the management of an organisation. I start here because an organisation such as the restaurant which I introduced in Chapter 3, whilst certainly much more complex than most mechanical systems, seeks to maintain as a constant its relationship to its environment. The restaurant desires a steady flow of customers. To achieve this it must maintain its reputation through managing its delivery of food and its ambience. It will, perhaps, on occasion change its style but, for the most part, if it is successful it will seek to maintain its designated style. In contrast, in the natural world an animal varies its relationship to its environment, even moment to moment, and therefore it is a more complex system to model. This I will explore in following chapters.

There are two inherent problems standing in the way of achieving the required stability and constancy in an organisation. The first is in knowing the current state in every part of the organisation; as a manager of an organisation, how do I know what is happening in every part of my organisation? It is not possible to watch all the people all the time! The *variety* of activities even in a small organisation of say 100 people is enormous. Here I use the word *variety* in the sense I defined in Chapter 3.

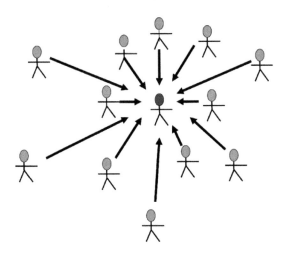

Figure 4.1: The problem of attenuation of information received

If I ask about the current situation in an email to my 100 subordinates – something easily possible – how do I deal with the 100 replies produced? By the time I have read, verified and considered the replies, the situation will have moved on and my replies will have become outdated. If I can deal with them at the average rate of, say, 6 minutes per reply, a day and a half will do it. But the time I have available is limited; I have other tasks to take care of, other roles to fulfil. At root I cannot know what is happening in all parts of my organisation, even one of this small size. The *variety* of activities is far greater that my capacity to take in the information required to have that knowledge.

The second problem is that of maintaining the coherence of my organisation. If I don't know what is happening, I must rely on the people within the organisation to carry out the tasks required of them – how am I going to persuade them to do that and make sure they do? Given the first problem, how can I possibly keep up? It is perhaps something of a surprise that organisations *do* hold together.

Figure 4.2 illustrates two situations where a less complex system seeks to persuade a more complex system to follow its will; the organisation seeks to persuade customers to buy its products, and the management seeks to persuade the other staff to carry out the tasks required to produce the product. Stating this in more general terms: the system must maintain its purpose and relationship to its environment and the steering mechanism must maintain a relationship to the system such that it can perform its purpose. But most disturbances to the smooth running of the organisation originate in the environment to which management has no direct access. The management must for the most part work at one stage removed and can only seek to influence the staff to overcome these disturbances. To draw on a simple analogy, this is like trying to reverse a vehicle which is towing two trailers one behind the other! I have seen that done but couldn't do it successfully myself.

One way that managers try to deal with these situations is by categorising, in order to reduce the variety they are confronted with; for example:

- in marketing, dealing with market segments not individual customers
- in schools, dealing with streams and classes not individual students
- in business, dealing with departments and not individual workers
- in finance, dealing with categories in charts of accounts rather than individual transactions.
- in life, judging people by some visible characteristic, or perhaps some particular exam results.

A second way used in management is not to deal with the variety managers are confronted with and/or to allow others to deal with it; for example:

- in supermarkets customers gather up, and even complete the transactions for, their own goods as opposed to the traditional way of serving customers
- in travelling, cars allow travellers to choose their own timing and route as opposed to buses and trains where timing/routes must be managed
- in crime, some crimes are not reported to the police, who are the official managers.

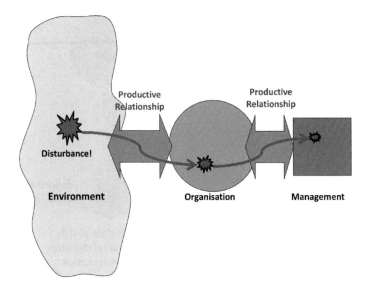

Figure 4.2 Managing the impossible

At the heart of these two problems is the concept of *variety*: the number of possible states of the system and its relation to the environment. All the examples above are concerned with reducing variety to manageable proportions. I can certainly reduce the variety I perceive if I shut the door to my office and ignore what happens on the outside, but this is unlikely to help any organisation.

Those familiar with the business world will know that in the management domain the issue of the advantages and disadvantages of centralisation versus decentralisation of management is a never-ending discussion. Large business organisations typically go through cycles of centralising and decentralising their management structures in order to try and get a grip on this problem. The central managers of a decentralised organisation, having left the management to others who act in ways that the central managers would not, feel that they must have more control and therefore begin a programme of centralisation. Managers of a centralised organisation eventually realise that they can't keep up with events and begin a program of decentralisation. But it is through the concept of *Variety* and *Variety Balance* within a *systems approach* that it is possible to get a grip on this problem. The essential problem is set out in Figure 4.3.

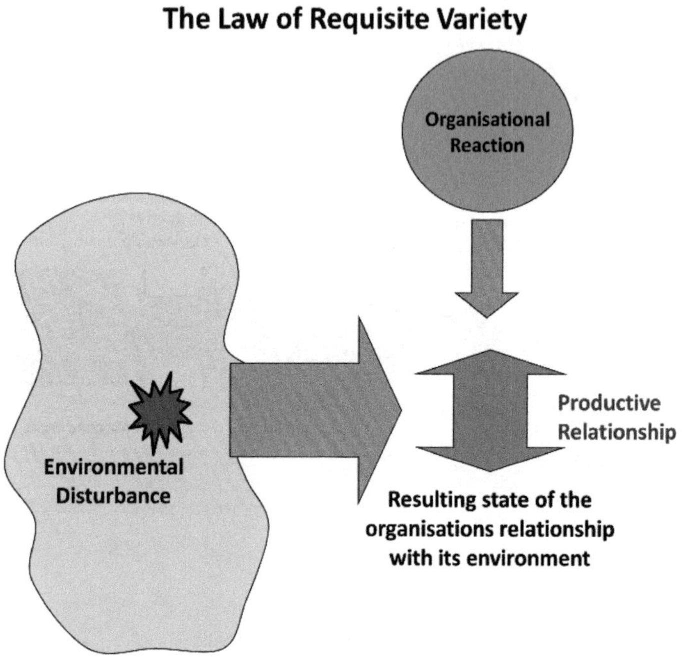

The Law of Requisite Variety

Figure 4.3: The steering problem

This was the problem first formulated by Ross Ashby (1956) who, using a mathematical approach, formulated what has become known as Ashby's Law. The law states that in order to maintain control, the variety of steering responses must be greater than or equal to the variety of disturbances that the system is subject to. So again I am faced with what appears at first sight to be an impossible situation.

I can envisage at least the start of formulating the necessary responses to environmental disturbances to my restaurant in the same way that Ashby formulated the proof of his law (Figure 4.4).

The list of possible disturbances will become very long indeed, much longer than the list of possible responses. The conclusion is that it is not feasible for one manager to know what is happening in their organisation, nor is it feasible for them to hold any particular purpose. So how does it happen that organisations hold together in a coherent way? One common way is to constrain the variety of the participants by threat, making it very clear that it is to the participants' advantage to carry out the tasks required. It was Machiavelli who famously said that it was better to be feared than loved. Slavery and bullying have a long tradition in management and of course also occur in the natural world.

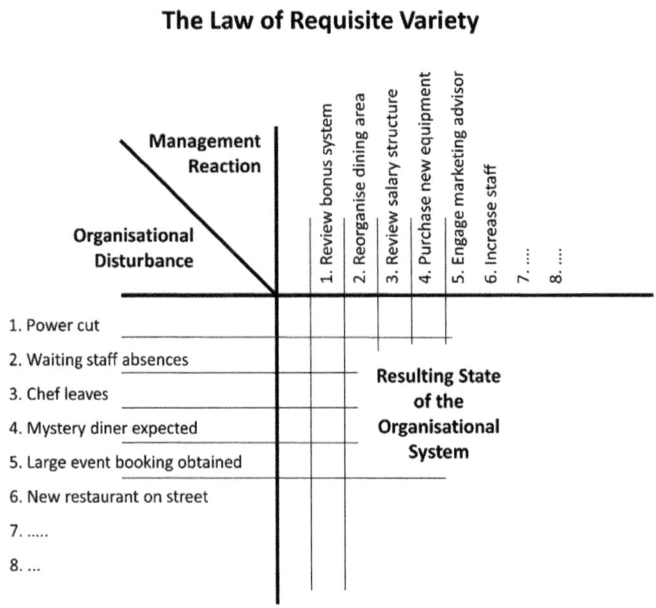

The Law of Requisite Variety

Figure 4.4: Responses to disturbances

This approach does achieve the purpose of constraining the variety of the workforce but having an unhappy workforce does not augur well for success. Sabotage and spying also have a long history.

4.3.   Dividing up an Organisation

There are more subtle ways of achieving the end that is required. The flaw in the thinking of the last section is to ignore the fractal nature of systems and processes (Section 3.6).

Organisations are always divided up into parts. Many different criteria are used, formally by function, by process, by product, by geography; and informally by building, neighbourhood, affinity of age, gender or interests; but what effect does one criterion rather than another have? The way we divide up an organisation will certainly have an effect on the relationships, the interconnections, between the various participants. How participants relate to one another determines the way the organisation functions. The communications that take place between people in an organisation are central to its ability to carry out its purpose.

In order to explore this I want now to examine the communication channel structure within an organisation – what Marx called "the relations of production"(Eagleton 2011). Figure 4.5 lists the number of potential communication channels between people within small organisations of various sizes. Starting with just two people and progressively adding an additional person easily verifies the pattern of the increase in the number of channels.

For any given size of organisation, we can see that the total number of potential channels is the number of people in the organisation multiplied by the number of people in the organisation minus one. Of course, in any organisation there are also channels from within the organisation to the outside environment but in consideration of the internal structure I will consider only the internal communication channels. Also, for the purposes of this illustration, I will consider a simple structure, but one which has the essential features of a typical organisation. This is enough to illustrate a general principle concerning communication and internal organisational boundaries.

Number of People	Number of Channels
2	2
3	6
4	12
5	20
6	30
..	..
..	..
100	9,900

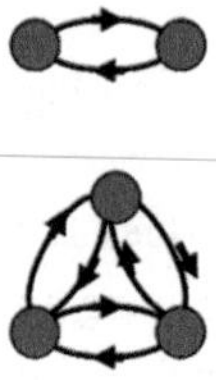

Figure 4.5: Dividing up an organisation; people and channels of communication

If I consider a small organisation of 100 people, then there are potentially 100 x 99 = 9,900 communication channels internal to the organisation. If I were to design this organisation, then I should consider the information that is to be carried in each of these channels to determine the structure I need. From my systems perspective, each person is a sub-system and the inputs and outputs for each person will have a specific purpose within the roles they perform. Designing this is a formidable task even in this case of only 100 people.

Since I wish only to get an idea of the complexity of this situation, for simplicity I will divide the organisation into five equal departments each headed by a manager: Figure 4.6. This gives a clear organisation chart. As normal I will require that management communication takes place between the managers concerning their tasks, issues and problems, and all other communication will take place within departments. I will organise departments by commonality of work, and place people in the same department close to each other in the same office and building to ensure good communication. In this departmentalised organisation we now have 5 x 4 = 20 management channels, that is those connecting the five managers, and 20 x 19 = 380 potential channels in each department. After introducing these internal divisions, the total number of potential channels of communication has reduced from 9,900 to just 1,920.

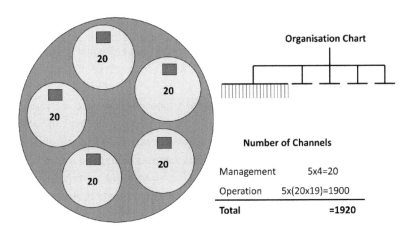

Figure 4.6: Setting up departments

For some organisations, departments of twenty would be too large, a span of control of about five being considered optimal. So, taking the argument a step further, I might wish to establish a further substructure within each of my departments and divide each into teams. For simplicity, again I establish in each department four teams of five people and propose that each team will be managed by a team leader. One of the teams will contain the departmental manager, so, for simplicity, I will designate to that person a dual role. I expect team leaders to communicate to discuss departmental business, and department heads to communicate as before, so establishing a normal hierarchical organisation chart:

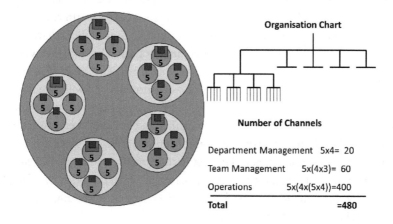

Organisation Chart

Number of Channels

Department Management 5x4= 20

Team Management   5x(4x3)= 60

Operations   5x(4x(5x4))=400

Total   **=480**

Figure 4.7: Setting up departments and teams

In this situation I have as before 5 x 4 = 20 departmental management channels. Within each department I have 4 x 3 = 12 team management channels, and within each team 5 x 4 = 20 operational channels. Restricting communication to these formal channels now gives a total of just 480 potential channels of communication. So, by introducing this structure within my organisation I have reduced the potential communication channels to less than five percent of the original 9,900. This seems a startling reduction.

However, although startling, it is an illustration of the universal principle that by introducing internal divisions within an organisation, most potential communication channels will not be utilised. In this case, in the formal structure I have removed nineteen out of every twenty channels. The need for informal channels of communication becomes clear. But it is also clear that I must be very careful how I introduce internal divisions within my organisation, if I am to avoid cutting vital links and preventing vital communication.

In establishing any organisation I must introduce internal divisions. First, people in my organisation in their roles will not just be communicating; there will be processes to be carried out. Time will limit the communications that can be achieved. Secondly, I must put people into buildings, and internal spaces within those buildings. The architecture of the buildings and the internal spaces will affect the structure of the communication flows as previously discussed when I drew the analogy with water flowing in a river. People will communicate more readily with others in the same office or building than with those at a distance. Information technology will partly alleviate this problem but not fully.

This illustration in structuring a small organisation shows that there are very good reasons why a common complaint in most organisations is that 'I often do not receive the information that I need to carry out my job'. Also, examining this small organisation illustrates a general problem of communication between parts of any human organisation, animal, or machine which must have a solution if the organisation, animal or machine is to exist as a coherent whole.

So how can we put internal boundaries within the organisation that minimise the damage to the communication structure of the organisation? The answer comes from continuing the train of argument I have established. Remember that each of the circles and squares of figures 4.6 and 4.7 represent processes; that is the central part of the input→process→output structure of a system. Then, remember, as noted in Chapter 3, that the basic systems model has a fractal structure. Therefore the systems view shown in Figure 3.3 can be applied to any sub-organisation of the whole. This was first illustrated with the example of a restaurant in Figure 3.10, reproduced here, at Figure 4.8:

(i)   Each sub-organisation should have a clear purpose resulting from a well-defined productive relationship with its (that is the sub-organisation's) environment

(ii) This purpose should minimise communications to other sub-organisations

(iii) Each of the sub-purposes should either reflect directly the purpose of the whole organisation, or in symbiosis with the other sub-purposes of the other sub-organisations add in an emergent way to the purpose of the whole organisation

If a sub-unit has a clear purpose and well-defined productive relationship with its environment, then that purpose can be understood and pursued with minimal need for communication with the sub-unit's environment. Although there may be many ways to divide up an organisation following these principles, in the natural world organisms have long achieved, through the mechanisms of evolution, effective and efficient structures; although not necessarily the *most* effective and efficient.

My conclusion is that the introduction of internal boundaries in any organisation will always reduce the possibilities of internal communication. Therefore internal boundaries need to be where communication is minimal. My primary task as a manager is to design appropriately the structure of the unit I manage. I must identify a set of purposes for sub-units that will work together to enable me to achieve my unit's overall purpose. Agreeing the purpose of each sub-unit with those who manage it will ensure that communication needs between units are minimised as far as possible. Then each of my sub-unit managers must do the same thing. A layered fractal structure will result when this is done through a whole organisation.

In principle, this analysis, driven by an analysis of purpose, can always be done in any organisation. Once the organisation is modelled in this layered way, a focus can be placed on any of these processes contained within a boundary. Each unit is a system, whatever level it appears at in the layered structure, and each of these systems can be analysed and modelled in the same way. In my experience, it is an unusual manager who pays attention to the structure of the organisation they manage. But whether a teacher managing a class of students, a parent managing a family or, at the other end of the scale, a chief executive officer managing a multinational corporation or a prime minister managing a country, structure matters. Organisational structures, architectural structures and information structures coded into technology systems, all contribute to the determination of who communicates with whom, when and how. The structure determines the ongoing dynamics of the organisation.

4.4.   Where is the Individual?

The restaurant that I introduced in the previous chapter exemplifies these principles. The purpose of the dining area system is to produce within its parameters the feeling of satisfaction and wellbeing through the dining experience. Many things contribute to this: the architectural ambience, the interaction with the table-waiting staff, and the experience of the food itself.

The focus is the customer and staff must reduce the enormous variety of attitudes and moods of entering customers to one of positive satisfaction and a wish to return. The kitchen processes a variety of raw food ingredients to the style of the restaurant, ensuring what leaves the kitchen for the dining area is of the standard required. If it is necessary to further divide the organisation, then each of the sub-units I identified needs to be considered in turn, and I can apply the same logic to each of them. Each sub-sub-unit exists as a system within the super-system of its surrounding sub-unit. The relationship is exactly like that between the sub-unit and the whole organisation. Therefore, I can simply use the theory again at the next lower level.

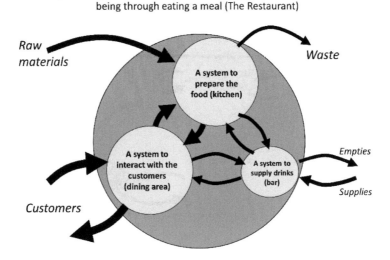

Figure 4.8: The beginnings of the fractal structure of a restaurant

In so doing, I build a system model of the whole organisation within which each of the sub-units exists as a system with its own clear purposes. Exploding this out into however many layers are necessary builds a system model pervaded and dominated completely by the whole system purpose and so the model is an holistic one. Within this model all the system parts – and their relationships – will flow from the analysis, so the model is a reductive model too. *Herein lies the power of modelling organisations from this viewpoint. This modelling process enables the retention of an holistic view in the pursuit of a reductive analysis.*

Taking a slightly different view, the analysis of the last section results in a conception of the restaurant organisation as illustrated in Figure 4.9. What could be considered individual, the restaurant, can also be viewed as several individuals interacting together at the next level of analysis. Even at three levels of analysis of a relatively simple system, the picture begins to look complex: a network of interconnected individual systems. In the biological world, the question of what is the individual arises in many situations.

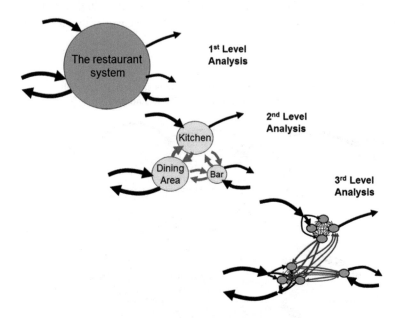

Figure 4.9: The Restaurant Analysis

Is the bee the individual or is the hive the individual? From this systemic viewpoint it is simply one, or the other, or even neither, depending upon the level of analysis. If I analyse further to lower levels then I come to roles that individual bees perform, or indeed, in the case of the restaurant, roles that individual employees undertake. Just as the screwdriver can take on different roles in relation to me; extant structures, an individual bee, or an individual employee, can take on different roles in relation to their environment. Evolution has made the cell an extraordinarily versatile structure which can take on many different roles with amendments to internal structures. Human beings too are also extremely versatile. I take on different roles as I contribute to different systems that I take part in, each of which can be analysed in this fractal way. Each role uses a set of mental models, the internal systemic structures, that enable me to carry out each of these roles. I am an individual in that I am an extant structure but a structure which can take on many individual roles and can contribute to many individual activities. In each of these roles my relationships with the others around me are special to that role, and it can be said that I am an individual but a different individual in each case. 'Individual' therefore is an artefact of the analysis: the purpose, context, and the level to which I take this analysis.

The analysis illustrated in Figure 4.9 is also important because it connects this approach to network theory and complexity theory. The diagram illustrates the way in which, as I pursue the analysis of any system down through the levels, what results is an interconnected set of nodes of an increasingly complex network. Network Theory and Complexity Theory (e.g. Cilliers 1998; Barabási 2003) are important systems disciplines, investigating many different systems but from a rather different perspective.

The approach I have taken shows already two important properties of networks: clustering and scale invariance. Each system process can be viewed as a cluster of nodes and built into the analysis is the fractal scale invariance that natural networks exhibit (*New Scientist*, 'The Borders of Order', 26 April 2014 pp.44-47). This answers the question that Barabási poses at the end of his book: why do we find clustered, scale-free networks in nature? The answer is that any living entity that consists of purposeful sub-systems can be analysed as a clustered, scale-free network. This is exactly the way in which living entities are constructed. I will discuss this further in this and later chapters.

4.5.   The Development of the Idea of 'Recursion'

Modelling, starting from a systemic base, necessarily leads to a fractal structure. This idea was first developed by Stafford Beer in a series of books developing his Viable System Model (VSM). This necessary layered or fractal structure of any organisation achieving viability came from a synthesis of logic, the control processes necessary in the management of an organisation and those necessary in the structure of the human nervous system (Beer 1959). Control situations are inherently complex because they involve parts of a system relating to each other as I have discussed. Relating means communications taking place between them, and these are *two-way feedback structures*. This is important because the dynamic behaviour of any complex situation is determined by the feedback structures. Stafford Beer came at the problem of the dynamics of complex organisations from an unusual viewpoint. His background was first an education in Western philosophy, followed immediately, due to the Second World War, by time in India where he studied Indian philosophy. After the war, returning to England, he soon became a senior manager in a steel company dealing with the practical problems of steel manufacturing. A major achievement for him in this time was to redesign the flow of material through the steel plant in which he worked in Sheffield, which enabled more than sufficient monetary saving to establish a 70-strong research department. This hands-on practical experience together with the understanding of Eastern philosophy gave him a rather different perspective to that of a typical manager trained in the West.

From his experience, Beer came to the conclusion that it is the way in which an organisation relates to its environment and the way in which the parts of any organisation relate to each other that determines how that organisation evolves through time. Thus, if any organisation is to survive and adapt to changing circumstances then it must pay attention to the way in which it relates to its environment and to the way in which its parts relate to each other.

Using these underlying ideas, Beer developed his Viable System Model as a model of what is necessary and sufficient for an organisation to be capable of adapting to a changing environment. In the first instance this was published in the first edition of *Brain of the Firm* in 1972. It was obtained by looking carefully at the structure and workings of the human body and its nervous system (Beer 1972/1981). From the understanding gained from that exercise he extracted what in his opinion were the logical necessities of organisation, communication and regulation which ensured, as far as possible, the survival of an individual human being. He proposed that these conclusions should apply to any system that must adapt to a continually changing environment.

Subsequently he developed and published a different derivation of the Viable System Model (Beer 1979) from the basic concepts of cybernetics.

Three fundamental concepts underpinned the model; the first, the idea of system that I introduced in Chapter 2, as a natural base on which to model complex situations. The second, the idea of variety, I introduced in Chapter 3 as fundamental to understanding how stability is achieved in a world of flux and change. The last, which Beer labelled *recursion,* is just the necessary fractal nature of any structure that is to remain stable in the world of flux and change, which I have developed in this chapter.

The second aspect of this recursive structure is a rather subtle but very important point. It is common to think in terms of managing an organisation – the important word here being '*an*' – implying that the organisation is singular – Figure 4.10. Indeed I can say that the management system identified in Figure 4.10 manages the restaurant. But looking at the diagram carefully shows that the management system actually manages in this case not one system but three (the kitchen – preparing food; the dining area – facilitating dining; and the bar – facilitating drinking) at the next lower fractal (or recursive) level. Beer argues (1959) that management can only be done from 'above' or 'outside' the system being managed. This argument is a logical one based on the famous mathematical work of Kurt Gödel – for more on this fascinating area, refer to Douglas Hofstadter's *Göedel, Esher, Bach* (1979). Thus, in the diagram I must differentiate the management system from the systems being managed since they are at two different levels. This is done in a diagram by the use of a rectangle rather than the circle, which is used to designate the managed system, as shown in Figure 4.10. Thus an important part of the recursive (fractal) structure is also the pattern of relationship between the management system and the systems (always plural) being managed. The Greek 'meta' is used to describe the relationship between the lower level systems and the higher level system; 'meta' meaning 'above', 'beyond', 'outside'. Thus the management system is a meta-systemic to the operational systems.

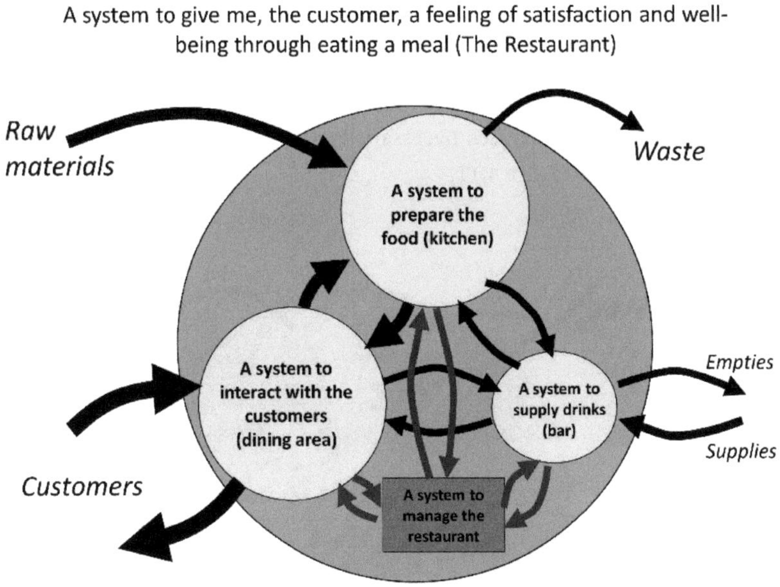

A system to give me, the customer, a feeling of satisfaction and well-being through eating a meal (The Restaurant)

Figure 4.10: Managing the Restaurant

4.6.   Steering the System

I have mentioned already the need for the table-waiting staff to manage the variety of customer attitudes and moods. In all but the smallest of restaurants it will not be possible for all of this management to be carried out by the owner: the variety of issues is too high. For the reasons discussed earlier (lack of knowledge and lack of oversight), management or steering in any organisation must take place at all levels in the fractal structure. In the restaurant, each staff member manages the task they are engaged in, the head chef and head waiter manage their respective domains, the owner manages the overall system.

The relationship of the restaurant to its environment is defined by the nature of the input and output flows of materials and information between the restaurant and its environment. Figure 4.9 illustrates, in the first level model, the way in which flows connecting the system to its environment appear as the primary concern of the restaurant owner. But as the analysis is pursued to lower levels it can be seen that these flows are in fact defined by the way in which sub-systems and sub-sub-systems operate. Therefore, whilst successful management occurs when the various tasks are differentiated appropriately in accordance with a fractal system structure, there must also be interconnection and coordination between levels. In outline, the steering processes necessary in the situation shown in Figure 4.9, are illustrated in Figure 4.11.

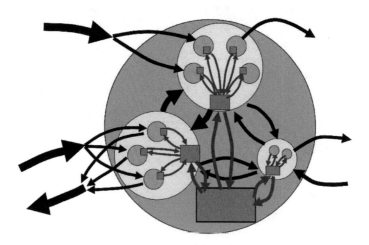

Figure 4.11: Steering the restaurant

Figure 4.11 shows three levels. It is important to notice that the pattern of management is the same for all systems, the whole system and each of the sub-systems, just as it would be for the sub-sub-systems if I continued my analysis to the fourth level. But not only is this pattern repeated but so is the pattern of relationship between adjacent levels. Each sub-system has its own purpose but those sub-system purposes must come together to achieve the purpose of the whole system. In any case, the whole system purpose will emerge from the interactions of the sub-systems whether designed in the way described above or not.

At this stage I can begin to see much more clearly the first of the processes hidden within what we generally call 'management'. In this I draw heavily on the discussion in Chapter 2. As a manager, I must first structure my organisation; that is, I must design the systemic sub-units and fix their purposes. I must do this in pursuing my management purpose as the overall manager of a unit with the best model that I can of the means to achieve that purpose.

In 1970 Roger Conant and Ross Ashby published a mathematical paper on control theory which concluded that every controller must contain a model of the situation being controlled, now known as the Conant-Ashby theorem. In Chapter 2 I illustrated how different models of the same situation result in different control actions: in that case a throat operation or performing a set of exercises. In this chapter I have mentioned the problem of different models of learning in education: the medieval traditional teacher expert broadcasting to passive students or the systemic learning from active experience (Whitehead 1929, Dewey 1938).

Different models of learning inevitably lead to different classroom processes and structures, and of course to different school management processes. In Chapter 2 I also mentioned that in slippery conditions a more extensive driving model is needed that includes the ability to handle a skidding car. This example highlights another aspect of the Conant-Ashby theorem: that the ability of the management is dependent on the quality of the models held.

In the case of the restaurant, my owner model may not be detailed enough or accurate enough to produce a successful design. I will need the knowledge and understanding of my head chef, head waiter and barman; i.e. I need their modelling input. As the restaurant owner, I manage three sub-processes *in conjunction* with the head waiter, head chef and barman. Their knowledge and understanding is essential to my structuring decisions. These structuring decisions are the first step in the creation of the internal environment within which the sub-systems will operate and, if carried out without adequate understanding, the sub-systems will not be able to achieve their potential. In any system, achieving the best management model in any complex situation will inevitably mean the involvement of sub-system managers.

Having designed what sub-systems are needed, I leave the detail of the processes inside these sub-systems to the sub-managers to design and control. I expect that they will take care of the variety within those sub-units and I will intervene only in exceptional circumstances when the purpose that I have set is not being achieved. The most important conclusion is contained in Figure 4.12.

Variety Reduction to Achieve Control

We achieve control by managing at a given level only the residual variety which is left unmanaged by the levels below.

Figure 4.12: The rule of steering

4.7. The Principles of Steering

I have spent most of this chapter discussing organisational management for two reasons; first it is an activity primarily concerned with holding a constant, that of a singular organisational purpose, in a world of flux and change. Secondly, it is where the ideas that underpin my approach have been most highly developed, particularly in the work of Stafford Beer developed in his Viable System model (VSM) (Beer 1972/1981, 1979, 1985). In the next chapter I begin the process of examining and modelling the management control system, firstly to release the constraint of considering only a constant organisational purpose, and secondly to begin to examine the necessary sub-systems of the managing control system.

In analysing the management control system Beer identifies five subsystems, which he labels System 2, 3, 3*, 4, and 5, as illustrated in figure 4.13. He explains that the choice of numerical labelling for these subsystems is as a result of the possible English words describing each purpose being inadequate. Each of the English words he examined as possibilities to describe the purpose of each of the subsystems carried a meaning which was subtly different from his conception of the purpose of the particular subsystem. I will discuss each of these subsystems in the following chapters: Systems 3 and 2 in chapter 5; Systems 5 and 4 in Chapter 6. All will reappear in discussing necessary subsystems of government in Chapter 8. It should be noted that as illustrated the fractal structure of the whole system implies that the set of these subsystems making up the managing control system appear at all levels in the managing control system for every operational subsystem at all levels.

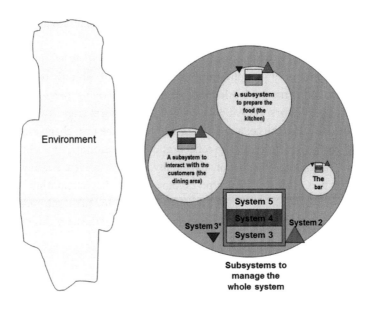

Figure 4.13: The restaurant identifying the subsystems of the management system

But, of course, in this chapter these discussions of management issues are examples, illustrating concepts, which can be applied more generally to all steering problems. It must always be remembered that the purpose of this writing is to suggest improved models of how animals (including human beings) navigate their environment in order to survive. The model used to enable survival is primary, and according to the Conant-Ashby theorem, the quality of the model is crucial in making the responses that enable survival. This chapter is a first step towards understanding the processes involved from a systemic viewpoint. Furthermore, since I am examining processes, a systemic viewpoint is the most effective and efficient one to take.

At first sight, environmental variety is overwhelming and survival is not possible; but not all disturbances are relevant. Billions of years of evolution have crafted viable living organisational forms for their ecological niche. Provided the rate of change of that niche does not exceed the organism's achievable rate of evolution, all is well. Organisms do not have to deal with the full variety of worldly disturbances. I have illustrated that, for a human organisation, a process/systemic approach aids understanding in how variety can be overcome and results in a fractal organisational structure. My understanding of biological and ecological structures is that evolution has resulted in fractal structures, being the natural way to ensure that environmental variety is controlled and constancy is achieved in a world of flux and change. I conclude from this that a process/systemic approach to modelling that world is an appropriate approach – and one that I now go on to explore.

Achieving control means managing at any level the variety left unmanaged by the level below – by implication, therefore, control can only be achieved bottom-up. In the natural world, pre-existing and surviving lower-level units come together to work in harmony synergistically to form larger units. Design is achieved through the collaboration of two levels. The next chapter explores first the nature of this necessary collaboration and, secondly, how change of purpose, which is one element of adaptability, can be achieved. For any whole there are two issues of concern: holding steady its relationship to the environment and holding steady the relationships between its sub-units. The second of these is in fact just an expression of the first at the next lower level.

5. Holding Steady: Control and Coherence

On the one hand physiologists have shown in a variety of ways how closely the brain resembles a machine: in its dependence on chemical reactions, in its dependence on the integrity of anatomical paths, and in the precision and determinateness with which its component parts act on one another. On the other hand, the psychologists and biologists have confirmed with full objectivity the layman's conviction that the living organism behaves typically in a purposeful and adaptive way. These two characteristics of the brain's behaviour have proved difficult to reconcile, and some workers have gone so far as to declare them incompatible.

Such a point of view will not be taken here. I hope to show that a system can be both mechanistic in nature and yet produce behaviour that is adaptive.

(W. Ross Ashby 1952/1960)

5.1.   Introduction

In Chapter 4 I discussed the initial stage of the systemic approach to modelling an entity that is capable of holding constant in a world of flux and change. The crucial property of a systemic model is that it has fractal structure. Implicit in fractal structure is that steering must occur at all levels but, as discussed, the steering function at any one level is meta-systemic to the sub-systems it steers: the steering function must be able to watch the sub-systems. In this chapter I want to discuss the detail of the steering mechanisms themselves and the relationship between the meta-systemic steering mechanisms and these sub-systems which are the focus of the steering mechanisms. The chapter naturally divides into three parts. The first is concerned with the basic steering mechanism, the cybernetic feedback loop. The second part builds on the exploration of structure in Chapter 4 and the necessary use of the cybernetic feedback loop in the maintenance of structural coherence of the whole system. From this point, the third part, I introduce further complexity. The assumption has been that the system being modelled has a constant purpose, constant relationship to its environment and no need to change that in order to survive. I now let go of that assumption. The third part introduces the cybernetic feedforward loop necessary to deal with a change of environmental circumstance necessitating a change of purpose and the selection of a different response model from the set of models available – in order to survive.

5.2.   Constraining Output

I need to return to cybernetics and the steersman of the ancient Greek ships. The responsibility of the steersman is to steer the ship to its ultimate destination. The ultimate destination comes from external considerations, the trading that the ship is involved in or perhaps the battle to be fought. The immediate and local course chosen by the steersman is determined by the external input and local conditions. The overarching reason for the decision choice comes not from the system itself but from outside the system under consideration, just as I would expect from the fractal considerations discussed in the last chapter. The management of this ship and the determination of the destination is part of the meta-systemic management of the fleet of ships of which this ship is a sub-system. Focusing on the ship itself, the question then arises: by what mechanism can the variety of possible courses be restricted to one particular course which achieves the end destination? To frame this in more general terms, what mechanism is needed to constrain a system to a required particular system state, or more likely a particular small group of states, out of the large variety of possible states that could pertain?

In Chapter 3 I introduced the steam engine and the governor as an example of two interacting systems coupled together by a negative feedback loop. That is a simpler example of steering than that of steering the ship, because all that is required of the governor is the regulation of the speed of the steam engine to that which is optimal for the load caused by the machine it is powering. Steering the ship involves many more factors including wind, waves and tide, and the possibility of future storms, all of which need to be considered by the steersman at any one time. The simpler example of the steam engine and governor allows me to show how the purpose of the governor is achieved, and from this example illustrate the necessities of steering.

The first of the necessities is to be able to sense changes in whatever it is that has to be regulated. As we saw in Chapter 2, this implies a way of perceiving what has to be regulated and a model to enable interpretation of the significance of what is perceived. If the steam engine is going to be useful as a source of power in driving machinery, the speed of the engine needs to be stable. As I described in Chapter 3 the engine is driven by using and releasing steam pressure.

Control is achieved by means of the *governor*, two balls fixed to a vertical rotating shaft such that they are able to move outwards, away from the shaft as the speed of the engine increases, and fall back inwards as the speed decreases. Therefore, the distance that the balls are from the shaft is a measure of the speed of the engine, and using this distance the speed of the engine can be sensed.

This governor mechanism is illustrated in Figure 5.1 where, as the speed of rotation of the shaft increases and the balls in rotation move out from the shaft, the rocker arm is pulled up. The linkage causes the rod connecting the governor to the engine to move to the left. As the speed decreases, the rod connecting the governor to the engine moves to the right. The mechanism enables the controller to watch the system under control. The governor senses the speed of the engine and can react to it through the connecting control rod.

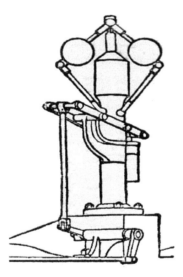

Figure 5.1: The Steam Engine Governor

The second of the necessities required is an *actuator*, which enables a change to be made to the operational system, the steam engine itself. The speed of the steam engine can be regulated by a valve which is able to vary the supply of steam, thus increasing or reducing the amount of power the engine is able to generate. The engine will normally be working against the load of the machinery it is driving so if the power reduces, the speed will reduce, and if the power increases, the speed will increase. In this case the movement of the connecting rod from the governor is used to regulate the steam supply.

The third and key component is a *comparator* which is able to compare the sensed value of the speed with the required value. What speed do we wish the engine to run at? This required value necessarily comes from the operating requirements of the machinery which the engine is driving and will be set by the human operator (i.e. from outside the system). In simple terms, the engine operator, by adjusting the length of the connecting control rod between the governor and valve controlling the steam supply to the engine, can set the required operational speed. This adjustment to the linkage will determine the amount of opening of the steam valve for a given position of the rotating balls, and so sets the relationship between the speed of rotation of the balls and the amount of steam arriving at the engine. In this case, the comparator is the human operator in conjunction with the connecting linkage.

Figure 5.2 illustrates the three essential requirements for controlling a process. First a sensor to sense some characteristic of the operational process that it is required to control. Secondly an actuator to change something in the process that will affect the characteristic that is under control, and thirdly a comparator wherein the desired value of the characteristic is compared to the actual value of the characteristic and an appropriate signal is sent to the actuator.

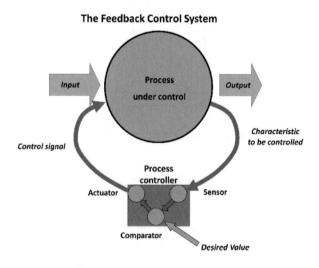

The Feedback Control System

Figure 5.2: The steering system as a feedback loop

This simple mechanical example enables the establishment of the necessities of a general model of steering. The three necessary elements arise directly from discussions in the last two chapters. To be able to steer in any situation requires a means of perception of those factors that need to be regulated. I can illustrate this with two more complex examples; first, an animal maintaining itself needs a means of perceiving food and predators. The requirements for this I discussed in Chapter 2 – the means to perceive and the models to interpret what is perceived. Second, human beings in all societies have expectations of how others will behave towards them and how they should behave towards others. This helps to maintain the stability of the society. But some break the rules. A means of regulation is needed in order to maintain stability. How are transgressions perceived, how would you describe the underlying model here, and what are the means of perception? I return to this question in Chapter 8.

The second steering requirement is that of being able to effect change in the system through some form of actuator. In the case of an animal the system that requires regulation in order for the animal to maintain itself is its position in its environment. It can regulate this by being mobile, moving towards a food source or away from a predator. In the case of the maintenance of a society the actuator is the means of eliminating transgressions of its rules.

As I discussed in Chapter 3, the need to maintain stability necessitates a negative feedback loop, and therefore the actuator will always act to decrease a perceived increase or increase a perceived decrease from some standard. This standard requires an input of a value from outside the immediate system: an input from a meta-system. It is important to note that this required value is not necessarily something precise as it would be in the example of the steam engine. Even a comparatively simple, and again mechanical, example of steering a vehicle along a roadway requires only that the vehicle stays within its lane going in the direction required. No-one drives to the precision of say exactly 1 metre from the edge of the lane. In most situations there is a variety of states that will satisfy the requirements of stability.

5.3. A Living Form in its Environment

In Chapter 2 I introduced the Craikian automaton as a way of illustrating how there are two possibilities of achieving the purpose of keeping the automaton on the table; i.e. surviving. The first of these possibilities is an example of how a very simple living form could control its environment. It simply reacts in recognising a particular circumstance. The automaton, on perceiving the edge of the table, turns. The three essential sub-systems of the control model are easily recognised: the sensor on the front; the comparator, the automaton's internal mechanism which enables the recognition of the edge; the actuator, the steering mechanism.

The second of the possibilities, more sophisticated and with an internal model, now looks rather interesting. The sensor is internal to the automaton but effects a change in the automaton's external behaviour. Not only that but, as I remarked in Chapter 2, an internal model gives the possibility of anticipating the change necessary, something I will come to in the third part of this chapter. It is easy to find examples of the use of internal triggers to change external behaviour.

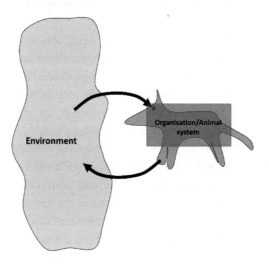

Figure 5.3: The Systems View: An Animal in its Environment

For a living form in its environment, regulation of food intake is driven by internal feelings of hunger and models of its home environment – a map of food availability likelihood. Avoidance of predators is driven by feelings of fear and again models of its home environment in relation to vulnerability – a map of threat likelihood. To its disadvantage, the dodo did not have a model of human beings as predator. The absence of an appropriate model means that the threat is not sensed and no control action can be taken. A system can only sense what its internal models enable. Feelings of hunger and fear come about through genetic learning; I will postpone for the moment discussion of how they relate to this control process, and I will leave you the reader to ponder the comparator for deciding on whether or not there has been a societal transgression in a particular case. This is a difficult question because what constitutes a transgression is not necessarily something that can be readily agreed.

For complex systems and complex situations, sensing and determining required values is itself problematic. Think of the difficulty of regulating the performance of a school in educating its students. Do exam results tell you very much about the effectiveness of the education a school provides? Can the changes in the mental models of the students be sensed sufficiently well by examinations at the particular point in time chosen? There is evidence that exam results are no guide to students' subsequent success in their career path. Successful entrepreneurs are often people who dropped out of formal schooling, so there must be a large question mark over the assumptions we make about exam results. In complex cases, the best sensing that can be done is to choose indicators that will stand for whatever is being controlled. The indicators chosen will, of course, depend on a model of the process, the mental model or the explicit model, that is used; i.e. the understanding accepted as that which pertains. In education, there are different models of how learning is achieved and which indicators are chosen depends on the model chosen. I will explore what a model of learning might look like from a systemic point of view in Chapter 6.

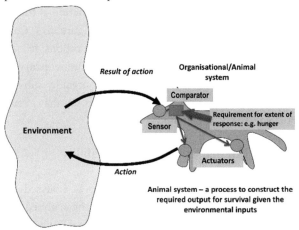

Figure 5.4: The Simplest Organisation/Living Form: unchanging purpose

The simple control model illustrated in Figure 5.4 is a model of the mechanism required to constrain a system, organisation or living form to a particular purpose/state or, perhaps more usually, a range of purposes/states. This mechanism is present in all our actions; for example, no explicit calculations are needed in catching a ball, just a sensor to sense the position of the ball in relation to me, a comparator to compare my position with the expected end position of the ball, and an actuator to change my position relative to the expected catching point. This is the first level of control, a feedback loop which ensures that any action is on course as intended; that is, during any activity, the control system's purpose is to maintain the purpose of the activity.

Ross Ashby, in his 1952 book *Design for a Brain,* introduces what he calls 'essential variables'. These are variables such as blood pressure, pulse rate and body temperature that describe aspects of an animal system. As he points out, the crucial aspect of essential variables is that they must be maintained within certain boundaries in order for the whole system to survive. In other words, they are variables describing aspects of the system that must be maintained, and to maintain them, control feedback loops must exist to constrain these essential variables to values that maintain the whole.

Similarly, with the management of the sub-systems in an organisation it is essential to constrain each of the sub-systems to their required purpose; for example, the kitchen, dining area and bar of the restaurant discussed in the previous chapter. These are the control feedback loops that ensure that the day-to-day functioning of the organisation continues as it should. In his analysis of managing systems, Beer designated this function as the purpose of what he labelled Systems 3 and 2, as illustrated in Figure 4.13. Constraining sub-systems to their designated purpose in organisations is a necessary function, but one which in nature is to a large extent unnecessary because in normal circumstances evolution has structured organic sub-systems to achieve a particular purpose. However this is not to say that in nature sub-systems cannot evolve to be repurposed.

5.4. Achieving Internal Coherence and Stability

In Chapter 4 I developed the idea that any system can be analysed into a set of purposeful sub-systems. Those sub-systems can then be analysed further into a set of sub-sub-systems and so on... developing a fractal model. In a dynamic world, in order for a whole system to remain a coherent stable whole, the activities of sub-systems must be constrained to those which produce that coherent stable whole system. In a dynamic world, the parts need to maintain the relationships among themselves which produce the whole. Families, communities or businesses must all maintain their internal cooperative and collaborative relationships to continue to function as those entities. In an animal, if the relationships between sub-systems are not maintained, perhaps because of disease, the animal dies. The maintenance requirements are critical: if they cannot be maintained the animal will not survive.

In a business management situation, as in the restaurant I considered earlier, I can set the purposes of the sub-units, but this does not ensure cohesion of the whole. A very common management error is to allow or even reward the optimisation of the performance of a sub-unit in isolation. But having sub-units pursuing their own optimisation separately does not lead to optimisation of the whole. One of my early experiences of management consultancy was to walk into what had been a successful kitchen unit manufacturing company which had collapsed after running out of money. Stored in every available space within the company premises were kitchen units at every stage of production. It seemed obvious to me, just walking in for the first time, that the production process had outpaced the ability of the company to sell its product. The purpose of the manufacturing sub-unit was being fulfilled effectively and efficiently but it had bankrupted the organisation. It seemed clear from that cursory look that the production of units should have been slowed to match the units sold, but the directors of the company had missed this obvious requirement.

A third mathematical theorem relevant to us here is the *Sub-optimisation Theorem* which states that optimisation of the whole is not achieved by separately optimising the parts; and therefore in this circumstance the whole system management must ensure good working relationships between sub-units. In any system there will always be a requirement for restrictions on the freedom of action of the sub-units in order to maintain cohesion of the whole. In other words, there must be feedback control loops whose purpose is to restrict the freedom of action of sub-units so that the coherence of the whole system can be maintained. I will now explore the nature of these.

5.5. The Relationship between Levels

The fractal layering illustrated in figures 4.10 and 4.11 is at each level a layering of systems working together to produce a super-system. In order that the sub-systems can work together there must be some common framework – at least flows of information and material leaving one sub-system of the super-system must be able to be received and understood by other sub-systems of the super-system. There must be at least common means of communication, and common models. In a human organisation this will take the form of things such as common understandings, a common language, a shared culture and material flows. So, for example, does the European Union, a super-system of a group of countries, have the necessary common framework? However, I can be sure that in a sustainable ecological system there *will* be a common framework – the relationships between all the living forms involved will give rise to the particular material flows in that ecosystem, for example a predator will have models of suitable and unsuitable prey.

Whatever commonality of shared structure there is will form a schema-system running through the sub-systems that form the super-system.

National government is one area where a systemic fractal approach to the establishment of a fractal governing structure would be of great benefit. A central government organisation cannot possibly deal with the variety of situations that exist in a population of millions of people. Just as in large corporations, in government there is also discussion around the centralisation or decentralisation of various functions. When decentralising, one of the most intractable problems in government is the relationship between different levels of government. For example, an examination of the complex levels of United Kingdom governance – central, regional and local government – doesn't give much of an indication of defining principles. There seem to be many variants with no universally agreed way of integrating different levels of government.

However, it is easy to find control or management situations where this problem is solved successfully. For example, there are many types of team game where there are sets of rules that apply to all the players taking part in a game, yet teams act autonomously within those rules.

This inter-level relationship in a multi-level model has exactly the fractal requirements necessary for any system. At the higher level of two adjacent levels there must be a frame within which all the sub-systems operate. Within that frame each sub-system can, up to a point, pursue its own purposes and establish its own additional culture and rules. The same relationship applies between any two adjacent levels. In governmental terms, the Universal Declaration of Human Rights and International Law is exactly a framework of this nature which should hold for all nation states, binding nation states to certain behaviours and, within that, allowing national expression of their own culture and laws. In the same way, Nation States' own national framework of law will apply to the whole nation, but similarly a nation can be divided into regions, each with its own culture and legal variations within the national legal framework, e.g Scotland within the United Kingdom.

A legal framework gives rights to citizens, 'freedom to' act and communicate in certain ways, but also provides constraints curtailing freedom, giving citizens 'freedom from' being on the receiving end of other particular acts and communications. A group can on its own account set higher standards of behaviour, but not lower, and it can certainly decide its own ways of communicating and acting within the common framework.

In any game, the relationship between the whole group rules – the framework that establishes the game – and the subgroup rules – the strategies that players use in taking part – and the way in which these two levels relate is an example of this. In a human organisation, these processes primarily involve discussion, which seeks to bring out commonalities between different cultures and adopting those as acceptable ways of working and being.

This process mimics the way in which nature builds complex structures. Over millions of years, single cell structures have come together and, when compatible (i.e. if there is a symbiotic match) multi-cell structures result. Nature builds from simple structure to complex structures. In nature, if the parts don't match there is no super-system. If the parts do match in some way there is a framework which works in such a way as to restrict the freedom of action of a sub-unit. But within this framework the sub-units have freedom of action, they are semi-autonomous; that is, they are able to work at achieving their purpose in their own way within that cohesive framework.

This is an essential part of managing the process of meeting external variety and, thereby, creating sustainability. Construction and maintenance of this framework is a crucial ongoing function for any organisation. The purpose of the framework is to ensure good working relationships between sub-units; they don't try to optimise their own performance to the detriment of other sub-units. The laws which keep most of us from violence, stealing and fraud are just such a framework ensuring the cohesion of a society.

The construction and maintenance of such a working framework involves both the understanding of the needs of the sub-units, and the meta–understanding of the way in which the units relate to one another. The role of the whole system management is to contribute the meta-understanding, exactly the role of a relationship counsellor, because there are aspects of any relationship that can only be perceived from the outside. However, a counsellor will not in general be able to perceive the internal processes of the systems they counsel. A counsellor cannot impose a framework but can facilitate the development of a framework. The framework (at any fractal level) is always a minimal agreed framework, just enough to ensure that the sub-systems, with differing purposes, work together to the common overall purpose at that level. We are all surrounded by these frameworks in our workplaces, in our families, in our sporting activities and in our lives generally. It is always instructive to identify the explicit and implicit rules of these frameworks, and also the levels within them, although we seldom do.

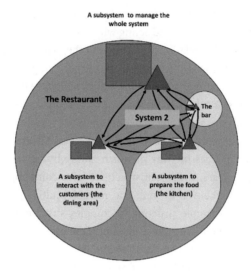

Figure 5.5: Ensuring coherence and stability – System 2

This systemic framework within which the sub-units operate has been labelled by Beer (1959, 1966, 1972/1981, 1979, 1985) as *System 2*, where *System 1* is the collection of sub-units themselves. As depicted in Figure 5.5, System 2 is an ongoing, systemic, causal conversation between the management of two adjacent levels in the fractal structure. Its purpose is to construct and maintain the framework within which the sub-units (System 1) operate. It is also important to realise that the framework at any particular level will be constructed within a framework emanating from the level above. These frameworks will produce a layered structure as illustrated in Figure 5.6. The overall framework which runs through the whole system is illustrated by the white colour. Each succeeding layer adding to that framework is illustrated by adding a little colour to that original. Thus each of the sub-systems at level 3 framed by the meta-systemic schema-system at level 4 can be different from others at that same level but, because of the framework commonalities, interaction is possible at each level as is needed to hold the overall system together.

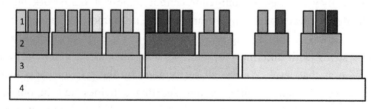

Figure 5.6: The layered System 2 Structure

98

Nevertheless, although very different local frameworks exist at the lowest level, they all exist within a common overall framework. This is illustrated at least metaphorically in Figure 5.6. The white background framework represents what is common to all sub-systems. Each sub-systemic level adds to that background framework stage by stage, level by level until, at the local lowest level, the local frameworks are very different from each other, represented by the changes in colour.

5.6.   Natural Frameworks

Examples of these frameworks are readily available. Elinor Ostrom (1990) relates one excellent example of a functioning System 2 framework, an empirical set of rules derived from working situations. Ostrom's example relates to an inshore fishery of 100 local fishers who operate in 2- or 3-person boats. Among themselves they developed a set of rules which allocated fishing sites to local fishers. The sites were allocated initially by lottery but the fishers move site each day which ensures that all have a chance at the best sites.

Ostrom reviews a number of such long-standing situations which she has researched and designates as Common Pool Resource (CPR) institutions and sets out the following set of design principles for such a framework.

1. Clearly defined boundaries
    Individuals or households who have rights to withdraw resource units from the CPR must be clearly defined, as must the boundaries of the CPR itself.

2. Congruence between appropriation and provision rules and local conditions.
    Appropriation rules restricting time, place, technology, and/or quantity of resource units are related to local conditions and to provision rules requiring labour, materials and/or money.

3. Collective – choice arrangements
    Most individuals affected by the operational rules can participate in modifying the operational rules.

4. Monitoring
    Monitors, who actively audit CPR conditions and appropriator behaviour, are accountable to the appropriators or are the appropriators.

5. Graduated sanctions
    Appropriators who violate operational rules are likely to be assessed [for] graduated sanctions (depending on the seriousness and context of the offence) by other appropriators, by officials accountable to these appropriators, or by both.

6. Conflict-resolution mechanisms
    Appropriators and their officials have rapid access to low-cost local arenas to resolve conflicts among appropriators or between appropriators and officials.

7. Minimal recognition of rights to organise
    The rights of appropriators to devise their own institutions are not challenged by external governmental authorities.

For CPRs that are part of larger systems:
8. Nested enterprises
    Appropriation, provision, monitoring, enforcement, conflict resolution and governance activities are organised in multiple layers of nested enterprises.

These design principles derived from empirical research provide an excellent guide to establishing a System 2 framework in a cooperative organisation, a System 2 which reflects the systemic considerations of this chapter for such a framework.

Another example from the natural world is that an individual ant or bee exists within the queen's pheromonal system.

5.7.   Changing Purpose

Thus far, I have restricted my discussion to systems with an unchanging purpose, illustrated by the example of the restaurant in an environment assumed to be unchanging. But some organisations change their purpose within their normal operations, for example a theatre company behaves differently in its rehearsal phase than it does in its performance phase as its purpose changes from perfecting a performance to presenting that performance. Ross Ashby (1952) was perhaps the first to formulate an approach to the requirement for an organisation to deal with a more complex changing environment. He considered the control responses of organisms rather than organisations. He considered, for example, how a kitten might respond to resting/sleeping by a burning fire, a burning fire having a variable heat output and on occasions burning material falling out towards the kitten.

He suggested that two levels of control are necessary, the first being the one I have explored – the negative feedback response to frequent low-level disturbances to the system. The kitten sleeps happily even though the heat output of the fire varies. The autonomic responses of the kitten keep its temperature within acceptable bounds. Second, he suggested that larger disturbances creating different environmental circumstances that overwhelm the normal negative feedback control would require a second level of control. If burning material approaches the kitten the autonomic responses are not sufficient and a different response is required; it will get up and move away from the fire.

Ashby's approach is based on the understanding that an organism will meet recurrent situations relevant to it during the course of its life. For the organism to remain viable it will develop responses to these recurrent situations through the processes of evolution. Thus an organism will contain inbuilt sets of response models that correspond to these different recurrent situations. The organism would then be able to call up a response model appropriate to its environmental circumstance and, if this circumstance changes, be able to recognise the change and respond appropriately. This is illustrated in Figure 5.7.

Changing purpose in this way is necessary for an organism not only in respect of disturbances due to changes in environmental circumstance, but also where the result of a previous action does not match expectation.

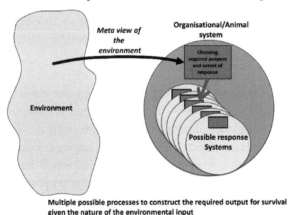

Multiple possible processes to construct the required output for survival given the nature of the environmental input

Figure 5.7 A system (organism) with multiple possible responses

This formulation of the way in which animals respond to a changing environment reflects what Gregory Bateson called 'zero learning' in the learning scheme set out in his seminal work *Steps to an Ecology of Mind*.

> Zero learning is characterised by specificity of response, which
> – right or wrong – is not subject to correction.
>
> (Bateson 1972, p.293)

Bateson points out that we describe situations such as 'I learn from the church clock that it is twelve o'clock' as learning. However, this understanding requires no change of structure of the nervous system, no change in any particular model. A signal is perceived which may be received as data, and simply noted, or may be received as information and cause action based on existing models. In this second case there is a change of model but not a change in any model; no structural change is required in the brain or nervous system. I reserve the word 'learning' to describe situations in which there is structural change. (I will explore how structural change may take place in Chapter 6.)

Figure 5.8 shows the result of putting together the two levels of control proposed by Ashby. First, the cybernetic feedback loop described earlier, and adding to it the cybernetic feedforward loop which anticipates the internal model requirement. The system to choose the required purpose of response in Figure 5.7 forming the process in the feedforward control loop now introduced was designated by Ashby the 'Γ gating mechanism'. In 1940 Jacques Monod discovered that in the presence of both lactose and sucrose the bacterium E.coli would first feed on glucose, and once this was exhausted, after a brief pause, switch to feeding on the lactose (Mukherjee 2016). This switch from one process to another, or in the presence of either lactose or glucose to switch on the appropriate feeding process is a simple example of an animal choosing an appropriate response to environmental conditions in the way illustrated in figures 5.7 and 5.8.

What is striking to me, and I am sure to those familiar with Beer's original VSM, is the similarity between this and the VSM model. But it should be noted that the VSM divides the purposes of a control system into those dealing with 'the inside and now' and those dealing with the 'outside and then'. This division omits the outside and now, which is essential in a changing environment. Beer developed his VSM with human organisations in mind which did not change their purpose, and his VSM did not consider changing purpose. In order to achieve homeostasis Ashby's view is that both feedback and feedforward control loops are necessary and an essential part of the control system.

Whilst there appears to be a direct correspondence between this modelling of the response system and Beer's modelling of it, which he labels System 3, this is not quite the case. Hence I propose that the VSM model can be extended to cover viable systems that are necessarily able to change purpose and have at their disposal multiple possible response systems with a reinterpretation of the purpose of Beer's 'System 3' to be Ashby's 'Γ' gating mechanism. This is a minor change from Beer's designation of the purpose of his System 3 to be 'resource bargaining' to the designation of 'resource allocation'. System 2 provides the control loops that hold the system together. System 1 contains the feedback control loops holding steady with respect to the environment (see Figure 5.8).

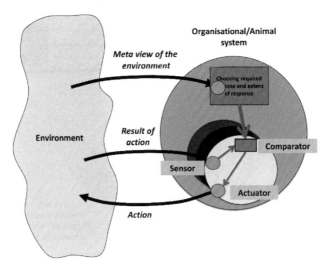

Figure 5.8: Multiple possible processes to construct the required output for survival given the environmental inputs

More generally, purpose is a fractal concept due to the fractal structure of inbuilt models (see Chapter 4). In most human and animal activities, purpose at some level is changing continuously. At the highest level I sit at the piano to play a piece of music. At the lowest level a muscle in my index finger contracts to produce a particular note, followed immediately by a change of purpose at some lower level to contract muscles which bring down my thumb. My inbuilt models produce the sequence of notes that fulfil the higher purpose of playing the chosen piece. This might be considered as a changing flow of signal from the reading of the notes from the sheet music through to a pathway from my eyes to my fingers, i.e. the natural sequence of response in that situation.

It is not possible to illustrate this dynamic switching flow in the simple static diagrams that I have drawn, but Figure 5.9 gives an illustration of such a situation, perhaps the flow to my index finger which produced that note. Before that note is produced my eyes have moved on to the next note and the new flow begins on the route to produce the next note. I leave it to you to imagine the way in which the broad black solid arrows illustrating the signal flow will change from moment to moment at various levels producing different actions from moment to moment. However this is still a rather simplified picture, since there is no explanation of how the management at each level occurs.

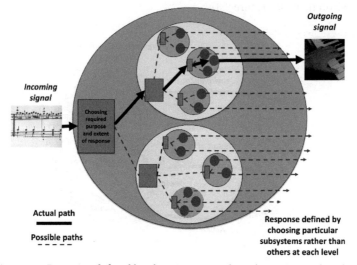

Figure 5.9: Response defined by choosing particular sub-systems rather than others at each level

6. Changing with the Flow: Learning

Plasticity, then, in the wide sense of the word, means the possession of a structure weak enough to yield to influence, but strong enough not to yield all at once. Each stable equilibrium in such a structure is marked by what we may call a new set of habits. Organic matter, especially nervous tissue, seems endowed with a very extraordinary degree of plasticity of this sort; so that we may without hesitation lay down as our first proposition the following, that *the phenomena of habit in living beings are due to the plasticity of the organic materials of which their bodies are composed.*

…the entire nervous system *is* nothing but a system of paths between sensory *terminus a quo [the point of departure]* and a muscular, glandular or other *terminus ad quem [the point of arrival]* The path once traversed by a nerve current might be expected to follow the law of most of the paths we know, and to be scooped out and made more permeable than before; and this ought to be repeated with each new passage of the current. …….. The most plausible views of the nerve current make it out to be the passage of some wave of rearrangement such as this. …….. If we call the path itself the 'organ' and the wave of rearrangement the 'function' then it is obviously a case for repeating the celebrated French formula '*La function fait l'organe*'.

(William James 1890, Vol I: 105-109)

6.1. Introduction

William James was one of the first to describe in a way that is coherent with modern neuroscience the ability of the brain and nervous system to change, due to both maturation and experience, and also to identify the systemic feedback nature of the change process, the basis of learning. The two foregoing chapters describe steering necessities for an animal to be able to navigate in a changing environment. But in those chapters it is assumed that the models at the heart of the living animal, the models of the environmental control system as far as the animal is concerned, are unchanging. But of course, in a changing environment, unchanging models may not be sufficient to ensure sustainability. In this perspective, a living entity survives by controlling the relevant aspects of its environment as far as is relevant to that living entity and as far as it can. In cybernetic terms, and from its perspective, the living system is the controller; the environment is the system under control. In order to remain viable in a changing environment and to survive in the longer term more complex living entities have the ability to change their models and also to develop new models. This ability is what I take to be meant by learning. Learning has the purpose of enabling changes to models to ensure that a living form continues to be able to survive in a changing environment so that it can as far as possible react to different situations including those it has never met before.

I distinguish between what I might call species learning and individual learning. Species learning, that is Darwinian evolution, is the result of changes introduced by the genetic and epigenetic processes in the transmission of models from generation to generation. If the change introduced results in the new living form having an improved matching with the changed environment, then that change propagates across the species due to the advantage gained. Species learning, thus, arises from genetic processes, operating generation to generation, and although the mechanisms are in general not the same as those of individual learning, the same principles underlie both. The same principles also underlie the learning which takes place in the immune system (Edelman 1987).

Individual learning is not required unless the environmental change that the living entity must deal with exceeds its capacity to deal with it in its reproductive lifetime. That is, in essence, short-living entities have no need of individual learning. Species learning is sufficient for these species to change with the changing environment; if neither process is sufficient for sustainability, then extinction results.

However, for a longer-living entity to be sustainable for the longer term requires that an individual be able to change its models and to develop new models. Individual learning – that is, the ability to develop, refine and change models in an individual's lifetime – has been accomplished in living entities principally by the evolution of the nervous system and brain. (Although, of course, the immune system is also a learning system.) The nervous system and brain of an animal is fashioned by the processes of maturation and the processes of living in a particular environment. What principally interests me here is modelling individual learning resulting from the ongoing processes of living and experiencing.

In the first section of this chapter I use the systemic techniques described in the previous chapters to develop a particular systemic model of learning. I also explore connections to other systemic models – those proposed by Gregory Bateson and David Kolb, Peter Checkland's Soft Systems

Methodology and William Perry's model of epistemological development. I also briefly examine neural mechanisms to ascertain whether the model developed is a viable model in the light of current understanding of the brain and nervous system.

6.2.   Systems Models of Learning: the context

In Chapter 2 I argued that perception and action result from stored models, within a living entity, which result in particular responses to particular patterns perceived by the living entity. I described the throat problem and the way in which the same input pattern in two different minds could produce two entirely different recommendations for action. The input→ process→output pattern of models identifies them as systems. In order to further understand this systems approach in Chapter 3 I developed the techniques of systemic modelling. In Chapter 4 this argument was taken forward in developing the means of analysing the structure of a system in a way that is both reductionist, reducing it to its parts, and holistic, retaining its coherence and interconnectivity. In Chapter 5 I then showed how, using this approach, an animal in its environment can be viewed as a system – a system capable of changing its purpose from moment to moment. Figure 5.3 illustrates the systemic view of an animal managing its environment. Figures 5.7 and 5.8 illustrate the way in which an animal chooses a particular response to a relevant environmental event. Further, I described the way in which this choosing is enacted by the subsystem of the managing system denoted by Beer (Figure 4.13) as System 3. In this chapter I turn my attention to the subsystem of the managing system denoted by Beer (Figure 4.13) as System 4.

To start building a model of learning I wish to set out what learning might entail in this modelling paradigm. It is important to remember that the models developed in the approach I am describing are at root non-linear. The first step to understand learning is to return to the river of Heraclitus, which appears static but does change its form over time because the flow of the water erodes the land and reshapes the riverbed. In turn the riverbed reshapes the flow of the water, so that there is a feedback loop between the shape of the bed and the flow of the water. This feedback loop is analogous to the loop that operates between a living species and its environment: the species and the environment coevolve, just as the flow of the river and the shape of the bed and bank coevolve. Again an analogous feedback loop (Figure 6.1) operates for an individual experiencing environmental changes that must be accommodated within its lifetime.

In the last few years neuroscientific research has shown that this feedback loop is essential for normal brain development (Denes 2016). There are windows in the maturation process of a developing child wherein interaction with the environment must take place for normal brain development to progress. Once the window closes, brain plasticity reduces and full competence is not achieved.

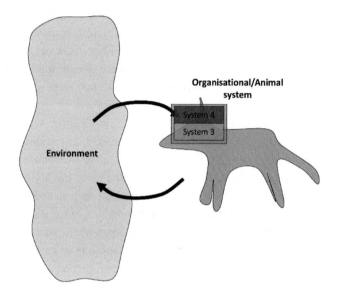

Figure 6.1: Subsystems of the managing system

6.3.   Changing a Model

The brain and nervous system of an animal consist of interconnected neurons. The base unit, the neuron, is a system: it receives inputs from its environment and transforms those inputs into outputs to its environment, just like any other biological cell. A neuron's purpose is to receive electrical inputs from a variety of other neurons, and if those inputs reach a threshold level, fire to distribute electrical output to a variety of other neurons. Neurons can be connected to up to 10,000 other neurons and can transmit a signal over a considerable distance within the nervous system and brain of an animal. Neurons form a complex interconnected web of systems, so that the electrical signals flow through this web of interconnected cells. The nervous system and brain are an enormously complex interconnected web of processes. This is exactly as I require for my modelling analysis shown in Figure 4.9 reproduced at Figure 6.2 with appropriate amendment to illustrate this point.

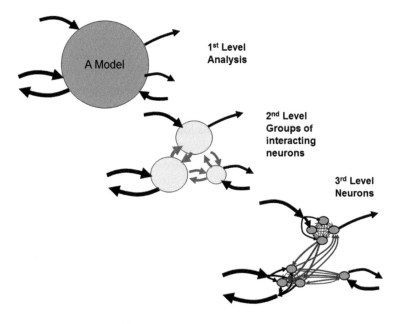

Figure 6.2: The modelling structure

In Figure 6.2 I identify a particular process. Say, for example, that I hear the rattle of the letterbox in my front door. The purpose of the process I then carry out is to retrieve the post, and my model at the first level of analysis represents that process. The second level analysis in the diagram represents the same process as I begin my analysis into sub-processes. This might consist of: get up, walk to the door, bend down, gather up the post, etc. In this simple case this is a sequence but in more complex models it is not necessarily so. The third level represents the sub-systems of each of these sub-system actions coded into the neuronal structures involved in each of those processes, the final level of analysis. The diagram is, of course, far from adequate in accurately diagramming such a process but does illustrate the point I wish to make, that these three diagrams are each representations of the same process at different levels of analysis.

At the level of the interconnected neurons, this structure was recognised by Paul Cilliers (1998) to be a complex system as envisaged in the science of complexity theory, and so fits exactly with the modelling techniques developed here. The structure of any neuronal network determines the relationship between the inputs to that network and the outputs from that network; models are encoded in the collective action of neurons. Further I can conclude that the purpose of the system that is an interconnected set of neurons is to process and store information in exactly the form I require for the modelling approach I have developed.

Donald Hebb (1949) was the first to propose that the strength of inter-neuronal connections was the origination of learning. Edelman (1987) described how these changes are driven by processes of maturation in the first instance and then by experiencing living in the animal's particular econiche. Interconnections that are not used decay, and those that *are* used, are strengthened. Furthermore, new interconnections come into being guided by those signal flows (Lettvin et al 1959). The neurons are arranged in neuronal groups of varying sizes (Edelman 1987), so perhaps I can identify the nervous system and brain as a recursively structured system, although a much more complex one than any I have described.

Bateson (1972) in the scheme that I mentioned in Chapter 5, defines a change in response to a given circumstance as Learning I, and gives a number of examples – breaking of a habit, the famous Pavlovian conditioning, rote learning and others. I continue in my efforts to learn to play the piano. When I started at around the age of fifty I was unable to move my fourth finger on either hand independently as was required. Now that is not a problem, that ability is achieved. I have slowly increased my ability to key the right notes and I am increasingly able to judge the changes that differences of pressure and speed make in the sound produced. I hear that I can play with increasing accuracy and speed, enabling variation in the way music sounds. As I practise, the model controlling my finger movements is gradually improved.

6.4.   Recognising Difference

I am slowly developing the ability in my playing to differentiate between the result a skilled person would produce and the variable result I actually produce. In order to achieve the improved result, I must refine both my perception and my action. This means that I must refine my abilities in order to differentiate between what are for me, very similar processes. The Law of Requisite Variety (Chapter 3) establishes the need for the variety of control responses of a cybernetic controller to be a least equal to the variety of possible disturbances to the system it controls. Developing an ability to be able to differentiate between similar situations in this way increases the variety of control responses exactly as required by Ashby's Law.

Figure 5.9 developed in Chapter 5, reproduced here as Figure 6.3, models the way in which a particular response is chosen from among many possibilities.

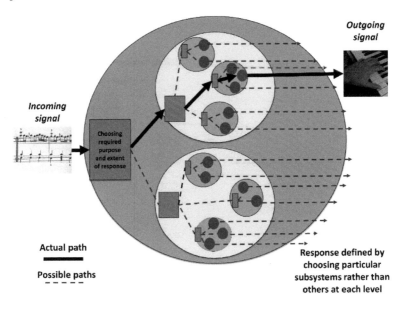

Figure 6.3: The response defined by choosing particular sub-systems rather than others at each level

Recasting this figure so that the various levels are separated out using the same colour scheme and including only the directly relevant pathways, I come to Figure 6.4. This illustrates one way of increasing the variety of potential command (see section 4.6) that can be envisaged: the ability to develop recognition and response that enable differentiation between similar incoming signals so that different responses can be produced.

As a first example of this process I come back to my problem of identifying birds. In this case, Figure 6.4 illustrates one model of bird being differentiated to two different models, perhaps chaffinch from bullfinch. Over the years with the aid of pictures and descriptions and much practice I have developed the ability to recognise the individuals of many species of birds, but I am still far from expert. A slightly different example of this process, where the level 2 system in Figure 6.4 must be reviewed, occurred in my youth, at a time when I was a very experienced cyclist. I bought a motorcycle. Because the activities were closely related but nevertheless different, at first I found great difficulty in balancing and controlling it. I had to differentiate the new experience from that of cycling.

Learning Process 1 - Constructing subsystems to enable the discrimination between two similar inputs.

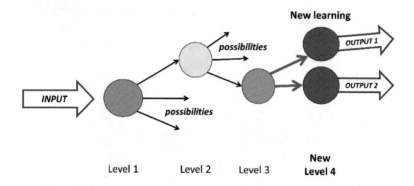

Figure 6.4: Recognising difference: Constructing sub-systems to enable discrimination between two similar inputs.

In the animal world I can also find examples of the way this kind of differentiation has developed. In Chapter 1 I considered the example of a frog which illustrated the necessity to differentiate out response models to hone the animal's ability to avoid a predator. Thus being able to differentiate out response sub-systems from an existing response system is perhaps the simplest way to extend response patterns to increase variety of potential command.

6.5.   Recognising Similarity

But what I feel is much more important and much more powerful is a second way of extending possible response patterns: that is to develop the ability to respond to new challenges, those that have never been met before. A living entity in a changing environment will naturally need to modify its responses as its environment changes, but will also meet challenges that have not been met before. This second requirement is a more powerful way of increasing the variety of possible control responses, of significantly increasing the variety of potential command.

In Chapter 5 I discussed the necessity for a living entity to be able to change purpose in response to different environmental circumstances. This required a living entity to contain sets of models for use in different situations. These sets of models are not necessarily connected. I suspect we have all experienced meeting someone out of their usual context and finding it difficult to identify them. Recognising the context and using the mechanisms of feedforward control we are able to choose appropriate response models depending on the environmental circumstance. I therefore propose that animals have the ability to review and compare models from different circumstances and extract from that comparison what is common. That common pattern of recognition and response can then be used to evaluate and react to new experiences that have commonalities of pattern with the original models. Piaget (1954) describes a number of examples of young children exhibiting multiple models prior to synthesis, for example:

> At 1 [year]; 3 [months and], (9) [days] Lucienne is in the garden with her mother. Then I arrive; she sees me come, smiles at me, therefore obviously recognises me (I am at a distance of about 1 metre 50). Her mother then asks her: "Where is papa?" Curiously enough, Lucienne immediately turns toward the window of my office where she is accustomed to seeing me and points in that direction. A moment later we repeat the experiment; she has just seen me 1 metre away from her, yet, when her mother pronounces my name, Lucienne again turns toward my office.
>
> Here it may be clearly seen that .... I give rise to two distinct behaviour patterns not synthesised......: "papa at his window" and "papa in the garden".

Piaget describes a process of learning derived from his examples that seems to me to align perfectly with the learning process implied by the modelling I describe here. I will return to the process he describes in Chapter 7.

As in the last section I start from Figure 6.3, reproduced from Figure 5.9. Separating out the various levels and including only the directly relevant pathways, I come to Figure 6.5, showing how the highest level in Figure 6.3 might have been formed. Hence, a second way of increasing the variety of potential command that can be envisaged is the construction of a meta-system from two existing systems. This meta-system will necessarily be a schema-system carrying the commonalties of the two (at least) systems from which it is constructed.

Learning Process 2 – Constructing a schema-system from the commonalities of different response systems.

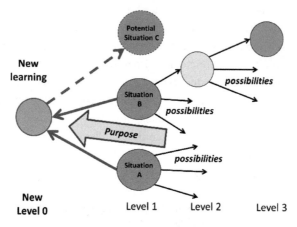

Figure 6.5: Recognising similarity: Constructing a schema-system from the commonalities of different response systems

The development of a schema-system in this way enables the development of new sub-systems and the concomitant increase in variety of potential command using the first method described. Beginning to play the pieces I practice on my piano competently enables me to play pieces that I have not yet met with some starting competence, and sometimes even to me startling competence. The technique of tackling a number of similar problems is often used in teaching, I have both experienced it and used it, introducing or being introduced to particular examples and then describing, or having described, the common pattern. This is a commonly used method in the teaching of mathematics. Being able to recognise a generic pattern is a powerful aid to learning, possibly even the root of creativity – being able to synthesise different experiences to create something new. Whilst perhaps this is something not easy to do, it is within the systemic paradigm something easy to understand and pursue. To be creative I need to seek different experiences in new circumstances, new contexts; my brain will take care of the rest.

The two learning processes taken together form a very simple but powerful learning mechanism, which increases the range and accuracy of recognition and response and permits the interpolation of new levels within existing recognition and response. A simple example of the last is that in my quest to be able to identify birds, once I have identified the commonalities of looks and behaviour between a buzzard and a red kite, I have also increased my ability to identify another bird of prey.

6.6.   Open Learning and Constrained Learning

> The evolution of ideas... is a tale of ever repeated
> differentiation, specialisation and reintegrations on a higher
> level; a progression from primordial unity through variety to
> more complex patterns of unity in variety.
>
> (Koestler 1970, p.227)

These two learning abilities, recognising difference and recognising similarity, which were modelled in the last two sections, are rather simple to describe in these systemic terms. In the first case I am developing the ability to differentiate between two similar processes where a singular recognition and response system is refined into two or more sub-systems. In the second case, I identify a schema common to two different recognition and response systems and then use that schema to guide reaction to a new situation. A meta-systemic schema-system is developed from two systems, from which new sub-systems can be developed using the first learning process. Both these possibilities increase the variety of responses available to environmental disturbance and therefore according to the Law of Requisite Variety increase the chance of survival.

Arthur Koestler in *The Act of Creation*, quoted above, describes many examples of situations in the progress of science where these two processes have been crucial in the development of improved models.

What has been really striking to me is that the development of a new meta-systemic schema-system is a powerful learning tool. In my youth when I studied mathematics, it was always a puzzle that one day a mathematical concept would be a complete mystery and the next day something completely obvious. The systemic model of the development of a new meta-systemic schema-system provides an explanation of how this could happen. Chapter 2 highlighted the notion that without a model there could be no understanding – this is the situation prior to the development of the meta-model. But then, through the synthesis of different experiences, a new schema-system is brought into being and, when it is used for the first time, there is an experience of sudden understanding where there was none before.

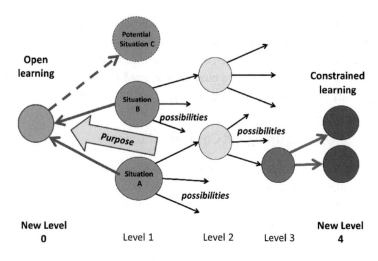

Figure 6.6: The two learning processes

I have designated these two learning processes *Open Learning* and *Constrained Learning*. Open Learning since the construction of a schema-system from two systems with similarities can be done in many ways; there may very well be a range of commonalities from which some are chosen as the basis for the new schema-system. The purpose for which the schema-system will be used will determine which of these commonalties of pattern are used. The existence of this meta-system will *open* new possibilities for the animal. On the other hand, the development of sub-systems to an existing system will be *constrained* to possibilities of recognition and action already experienced but now refined.

When I first came to this understanding I was struck by its alignment with the work of George Lakoff (2008) on the power of framing. Lakoff proposes that conservative thought assumes, is framed in, a hierarchy of authority and is founded on the notion that morality is obedience to that authority. For a conservative the authority is assumed to be legitimate, inherently good, knowing right from wrong, and functioning to protect us. Obedience to legitimate authority requires both personal responsibility and discipline, and a propensity to be loyal to the authority. Punishment and force are used to reduce the inherent proliferation of variety. Freedom is seen as a functioning within such an order. Anyone is free to act as long as they act within the rules laid down. Clearly, within such a system all learning is constrained. Accepting unquestioningly a particular frame allows only the possibility of elaborating response sub-systems within that frame. This is Constrained Learning.

On the other hand, Lakoff proposes that progressive thought is built on caring and empathy together with the responsibility and strength to act on that. Having empathy entails not requiring conformance within a particular frame but being open to learning from others and *their* way of living. He notes that enabling empathy and care requires that government have two intertwined roles – protection and empowerment – and that the role of progressive government is to maximise freedom for all to enable learning from others and understanding others and their viewpoints. This can only be achieved from the understanding that government has to be built from the bottom-up in a fractal way as outlined in Chapter 5. Within a framework of protection and empowerment people can come together to interact with each other despite their differences. If they do this they will, through comparing and contrasting their own ways of living with other ways of living, learn of the commonalities through the process of Open Learning.

I can conclude that it is not surprising that education systems that promote Open Learning (for example, in Finland) achieve much better results than those which favour Constrained Learning (for example, that of the UK).

6.7.   Systemic Models of Open Learning: The Learning Cycle and Soft Systems Methodology

Two strands of thinking have produced models that provide a heuristic to achieve Open Learning. David Kolb's (1984) follows the tradition of those who claim that learning arises most effectively from experience. This tradition includes Alfred North Whitehead (1929) and John Dewey (1938). The Kolb Learning Cycle, Figure 6.7, proposed in the mid-1970s, is systemic in general terms.

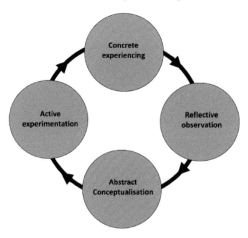

Figure 6.7: The Kolb Learning Cycle

However it is superseded by the explicitly systemic Soft Systems Methodology (SSM) (Checkland 1981), Figure 6.8, which has been a widely used heuristic since it was published.

Both follow a very similar cyclical heuristic as illustrated in Figures 6.7 and 6.8. Stages 1 and 2 of SSM correspond to Kolb's *Reflective Observation* but delve more deeply and perhaps more formally into the situation under consideration. Stage 1 is thought of as a wide investigation into the problem situation which should include all those involved. Checkland specifically enjoins us not to focus on a 'problem' but to also consider its situational context, that is to consider the whole problem situation. Stage 2 is the recording of the findings of this investigation in a *Rich Picture*. The Rich Picture consists of anything that contributes to the understanding of the situation under investigation gathered into one place. This is often just a summary picture with cartoon representations of various aspects of the situation, but in principle can be much wider than this. It is important to capture multiple perspectives and understanding of interconnections in this process, and also important not to exclude factors by assuming that 'the problem' is understood.

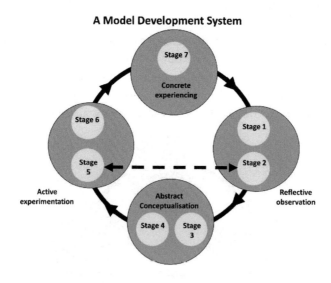

Figure 6.8: Soft Systems Methodology

Stages 3 and 4 of SSM correspond to Kolb's *Abstract Conceptualisation* but connect directly to the modelling techniques that I have explored in previous chapters. At Stage 3 the methodology directs us to distil from the Rich Picture the purpose or purposes of relevant systems in the problem situation (in Checkland's terminology, *Root Definitions* of these relevant systems). The relevant systems are exactly as I defined in Chapter 3.

And in Stage 4 we are directed to construct the conceptual model of those systems as I have described.

Stages 5 and 6 of SSM can be thought of as Kolb's *Active Experimentation*. In Stage 5 we are directed to explore the comparison of the use of the models developed with the results of the initial exploration recorded at Stage 2, perhaps to modify and improve the systems models constructed. The analysis of the use of the systems model or models, Stage 6, will give rise to recommendations for action in the problem situation. Thus learning emerges at Stage 6, and action to improve the problem situation based on this learning is to be taken at Stage 7.

Perhaps the most important aspect of SSM is that when a group of participants in a problematic situation undertakes the process it is unlikely that all those involved will see the situation in the same way, and therefore agree on a single result at the end of Stage 4. The encouragement to consider different systems at Stages 3 and 4, perhaps from different perspectives taken by the different participants in the problem situation, provides an opportunity for facilitating open learning so that new thinking can take place. I will return to the importance of perspective in the next chapter.

6.8.   The Making of a Mind

We start life with human perception and a developing human brain and nervous system, the blueprint of which has been developed over many thousands of years by the processes of evolution. We are born into a particular society and develop into a member of that society. As Raoul Martinez (2016) writes:

> Our entry into this world may be arbitrary, but the world that greets us is not. Numerous forces vie for our attention and loyalty. Our minds are a battleground for competing ideas. The outcome of this battle determines who we become and the society we create… Over the course of human history, countless people have been conditioned to defend oppressive ideologies, support destructive regimes and believe downright lies.

There are many theories of learning but the work of William Perry (1999) on epistemological development stands out as a set of ideas which describes in general terms how a mind develops. Perry proposed a 9-stage model of epistemological development as shown in Figure 6.9. A child is born into a relatively closed world of a family and absorbs the culture and attitudes of that family. In the absence of any alternative views these values are taken as absolute. As a child grows and expands its range of interactions with people other than its family, it meets alternative views. It can either reject those alternative views outright as wrong, or perhaps, if they are met in a new context, store and use them in that new context. Perhaps, in the first instance, the child will not recognise the inherent contradictions, but create alternative models for consideration later. In Perry's words:

> In Positions 1, 2, and 3, a person modifies an absolutist right-wrong outlook to make room, in some minimal way, for that simple pluralism we have called Multiplicity.

Once alternative models are stored, as I explored in the previous chapter, the mechanisms of Open Learning become available for developing more sophisticated generic views.

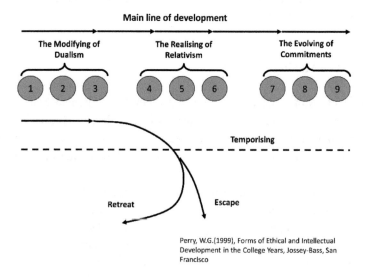

Perry, W.G.(1999), Forms of Ethical and Intellectual Development in the College Years, Jossey-Bass, San Francisco

Figure 6.9: The Perry developmental scheme

This is a process which then continues. I remember vividly the many times my beliefs were challenged by the move from living for twenty-five years in Britain to living for two years in the United States. I was forced to understand that many of my beliefs based in the culture that I grew up in were not the beliefs of those who grew up in the culture of the United States. Some people reject and retreat from those new experiences, putting off dealing with the new experience, escaping from a situation that is challenging or simply retreating back within a comfort zone. I can think of people who live in a country other than their home country but only mix with those like them. In Perry's words:

> The positions of deflection (Temporizing, Escape, and Retreat)
> offer alternatives at critical points in the development.

But, being of an enquiring mind, I did not. Rather, I sought out experiences of other cultures and have been reading widely throughout my life embracing the complexity of the world that I have found. But in all this exploration I have to choose which of these ideas and behaviours I wish to adopt in any context in which I find myself. Again in Perry's words,

> In Positions 4, 5, and 6, a person accords the diversity of
> human outlook its full problematic stature, next transmutes
> the simple pluralism of Multiplicity into contextual Relativism,
> and then comes to foresee the necessity of personal
> Commitment in a relativistic world. Positions 7, 8, and 9 then
> trace the development of Commitments in the person's actual
> experience.

Each transition is characterised by a meta-systemic leap as set out in section 6.3, a leap of open learning from a more simple view of the world outside to a more complex view of the world outside. The pattern of development that Perry describes is one of continual Open Learning, unless that pattern of Open Learning ceases for fear of the broad open spaces of the intellect, the complexity of the world out there, or any other reason to remain secure in the familiar. In summary, in Perry's words:

> Most broadly, the development may be conceived in two major
> parts centering on Position 5. The outlook of Position 5 is that
> in which a person perceives man's knowledge and values as

relative, contingent, and contextual. The sequence of structures preceding this Position describes a person's development from a dualistic absolutism and toward this acceptance of generalized relativism. The sequence following this Position describes a person's subsequent development in orienting himself in a relativistic world through the activity of personal Commitment.

6.9.   Signal Flows and States

In the final section of this chapter I would like to consider briefly how the nervous system/brain of an animal might function in a way that aligns with the modelling approach I have developed. What writers on neuroscience tell me is that the nervous system/brain consists of a complex interconnected web of neurons that live within a body. Candace Pert (1998) describes the way in which the nervous system communicates and functions within the meta-systems of the immune system and the endocrine system. Because the neuronal interconnections are strengthened by use and die away with non-use I deduce that there is a feedback loop between the flow of signals and the structure of the network. This structure of the network of neurons is a fractal structure as I explored in Chapter 4 (Werner Undated). The structure is not a simple linear structure (as perhaps my rather simplified example might indicate) but is a structure that contains feedback loops at all fractal levels (McCullough 1965).

Perhaps the starting point here, as in Chapter 2, is that the brain is a pattern recognition system, which is able to differentiate between different incoming patterns and produce response signals appropriate to a particular incoming pattern. In that simplest of situations, a signal pattern flows from the point of receipt to the point where appropriate actions are produced. In more complex situations, often when watching response patterns, it looks as if an internal tape is being played through.

For me the crucial understanding is that signals flow through the system. I propose that as a flow pulse moves along a bundle of nerve fibres its spatial cross section and pulse shape will have a pattern related to the initiating source or sources. This pattern is analogous to the way in which a pulse moves along a waveguide or optical fibre. In an optical fibre, information is transmitted in exactly the manner required. The great advantages of the way in which the brain and nervous system is constructed are, first, that energy is fed into the system throughout, so signals can be maintained. Secondly, at any point in a cross section or in the length of the bundle of nerve fibres, a signal can be extracted. So that the precise cross-sectional shape of the pulse can be detected and information extracted.

Thirdly, through this mechanism, parts of the cross section of the pulse can be steered in a particular direction at each point. In a simple network it is not necessary for any active steering to take place, the spatial pattern across the flow of that flow interacting with the neuronal pattern of complex system structure will enable flow patterns to be discriminated one from another. It is as if the flow starts from the top of a systemic structure and moves down through the levels choosing a particular path at each management juncture, and thereby choosing the appropriate operational path. However, in a more complex network containing higher level feedback loops, the signal flow can be maintained and can easily seem like the playing of an action tape when viewed from the outside.

Having a single flow of signal is of course a rather simplistic picture; in a real brain the inputs come from a set of senses. Several signal flows are simultaneously processed and combined. Outputs are also multiple, but this way of approaching the modelling of brain function seems to me to bring better understanding. Neuroscience research is progressing very fast but still has not achieved an understanding of the processes that take place. I believe that this systemic approach would enable better understanding.

Thinking about the learning mechanisms proposed in this Chapter we know that,

> ... humans continue to produce fresh neurons in the
> hippocampus at a steady rate well into old age, adding about
> 1,400 per day.
>
> (Kheirbek & Hen 2014)

Tests also show that inhibiting new neuron production reduces the ability to differentiate between similar situations. Both learning processes require brain changes. New neurons and new neuronal connections provide for the possibility of learning. Constrained Learning requires greater density of neuronal interconnections in order to differentiate between similar flow patterns. Open Learning requires the comparison of different flow patterns and the reinforcing of commonality. This is a process that depends on feedback. There is clearly much more to learn about how these processes function in a brain and how they can be modelled.

7. Recognising the Importance of Perspective

If actually all our knowledge is derived from perception, to say that the things 'really exist' has no meaning; for if we have 'perceptions of things' it makes no difference whether the thus perceived things exist or not. Hence 'to be perceived' is equivalent to 'to exist'.

(Werner Heisenberg 1958)

The study of sensorimotor or practical intelligence in the first two years of development has taught us how the child, at first directly assimilating the external environment to his own activity, later, in order to extend this assimilation, forms an increasing number of schemata which are both more mobile and better able to inter-coordinate. … The more numerous the links that are established among the schemata of assimilation, the less it remains centred on the subjectivity of the assimilating subject, in order to become actual comprehension and deduction. …The increasing coherence of the schemata thus parallels the formation of a world of objects and spatial relationships, in short, the elaboration of a solid and permanent universe.

(Jean Piaget 1954)

7.1 Introduction

Heisenberg states above that we learn from what we perceive, and Piaget describes how the world of the infant develops from its subjective experience to construct a world of objects. In this chapter I will explore the relationship between WEIRD thinking, the world of objects, and the systemic approach I have taken to modelling and understanding. I will show that perspective plays a key role in their relationship. The idea of *perspective* has played an important role throughout the previous chapters. In Chapter 2 I asked you to imagine the perspective of a tick managing its environment. In Chapter 3 I discussed the way in which I could knowingly take different perspectives on a particular organisation by attributing a particular purpose to that organisation. This purpose could be taken as a starting point for modelling the organisation as a system and would enable thinking precisely about the relationship it would have to its environment. In Chapter 6, perspective again plays an important role in learning. In particular I described Soft Systems Methodology which was developed to enable the bringing together of the different perspectives of members of a group, learning together from each other, and thereby to be able to agree on action to improve their situation.

The word *perspective* carries several meanings. I wish to highlight just two; visually, it can denote viewing a scene or an object from a particular spatial point: a geometric interpretation of the word perspective. But in addition, as I described In Chapter 2, because perception relies on pre-existing mental models, two people experiencing the same complex situation will not necessarily perceive that situation in the same way. As a result of their own unique life experience each will have their own *perspective* on that situation because their mindset restricts what they can perceive. But each of these uses of the word *perspective* denotes that the incoming signals *perceived* by an individual are a restricted subset of the total of the signals it is possible to perceive emanating from the situation, and it defines the relationship between what is perceived and the perceptor.

To explore the relationship between WEIRD thinking and my approach in the previous chapters I will first describe the development of each and seek to define each sufficiently for the relationship between the two to be clear.

7.2 WEIRD Thinking

In the Western world we have developed an accepted way of thinking about thinking that frames the world in terms of *objects* characterised by *attributes* which Henrich et al describe as WEIRD thinking or as *analytic reasoning*:

> Analytic thought involves a detachment of objects from contexts,
> a tendency to focus on objects' attributes, and a preference for
> using categorical rules to explain and predict behaviour.
>
> (Henrich et al 2010)

This way of thinking dominates and inhibits our thinking because we focus on the object; any relationships that the object has to its environment move into the background. Because relationships are not forefront, we have seen ourselves as separate and apart from the natural world. In recent years this has been sufficiently challenged for us to realise that this way of thinking may have already guaranteed *Homo sapiens*'s own early evolutionary extinction.

We have not paid attention to many of our links with the natural world, importantly and particularly those around what we view as waste, the volumes of which are now so great as to impact on our own survival. We are discharging waste into the natural world beyond its capacity to deal with it.

Many writers trace the roots of WEIRD thinking to the ancient Greek philosopher Parmenides, as I did in Chapter 1. But it was the developments in the 17th century in science and philosophy that really established the dominance of this approach. Science then was the study of material bodies and their behaviour, with a clear separation from the study of life and living. The difference between the animate and the inanimate was thought to exist because the animate contained a life force. This concept can be traced to the thinking of Aristotle (Aristotle 1986) and was still the dominant thinking in the 17th century. There was an explosion of scientific advances at that time which began with Galileo who began the systematic use of observation and experiment to study the inanimate world. Galileo's formulation of the fundamental laws of dynamics, that the acceleration of a body is proportional to the force acting upon it, began the process of using mathematics to model our experience of the world around. This success in the use of mathematical modelling was then extended by Newton with great success in dynamics, light, and the motion of planets and gravity. Significant also was the success of Newton and Leibniz in developing calculus to solve the problem of how to introduce mathematical models able to explain movement into the static world of objects. Up to that time, Zeno's paradox, that to travel from A to B you had first to cover half the distance, then again half the remaining distance, and so on, ad infinitum, was unsolved. Covering infinitely many distances in a finite time is not possible, but experience tells us that it is indeed possible, but this paradox had had no satisfactory solution until the development of calculus. The success of Galileo and Newton, followed by many others in Europe realising the importance of observation and experiment, produced the successful modelling which began the explosive success of WEIRD thinking in modelling phenomena of the natural inanimate world.

The philosophical justification began with Descartes in his search for certainty as a foundation for his understanding of understanding. He sought a starting point independent of the authorities of the ancient world and what is most famously remembered is '*Cogito ergo sum*' – I think therefore I am. But where do thoughts come from? Descartes' solution was that the mind was of a completely different nature to material bodies, and thoughts came from God. Since God, being perfect, would not dissemble, thoughts were about what really existed. Descartes thereby established a justification for the belief that what is perceived and thought is what exists in the world outside. This approach fitted the European culture and beliefs of the time, and also fitted with the developing scientific understanding of our experience of inanimate objects.

WEIRD thinking is based on the idea of object, attribute and mechanism as formulated by Newton and others and further developed within the world of physics. It is based within a belief that this world of objects is as it is perceived by each of us, what we see is what is really there and we should automatically agree on that. In philosophical terms this belief can generally be termed *naïve* or *direct realism.*

In order to differentiate illusions from experience of the '*real world*'' it is accepted that those experiences are private sense data, and other explanations are sought, for example in the supernatural.

The belief in this '*real world*'' gave rise to the idea of knowledge detached and independent of the knower. Thus '*objective*' has come to be an adjective in description of a situation, an opinion, or reportage which carries the meaning that this is the 'correct' or 'real' view that all should believe, independent of any particular observer. It is generally a view that informs us directly of things and their attributes, assumed to be a description of the thing as it is rather than what we would observe in the particular circumstances in which we were looking. WEIRD thinking accounts for the geometrical sense of *perspective* perfectly adequately in that we understand that objects that have spatial extent can appear in various forms due to distance or their 3-dimensional shape.

As a result of the success of this approach in physics, other disciplines, including biology, psychology and economics, also adopted this worldview. From there this mechanistic worldview became dominant in the Western world, so much so that even today it can be asserted that:

> We all have the intuition that the world is made of rigid
> objects. In reality, it is made up of atoms, but at the scale of
> which we live, these atoms are often packed together into
> coherent entities that move as a single blob and sometimes
> collide without losing their cohesiveness..... These large
> bundles of atoms are what we call "objects." The existence of
> objects is a fundamental property of our environment. Is this
> something that we need to learn? No.
>
> (Dahaene 2020)

But all this has never been without challenge. Reading on you will find that the answer to the question Dahaene poses above – "Is this something we need to learn?" – is reversed to 'Yes' in the approach I have taken.

7.3 Challenges to WEIRD thinking

The enormous success of the Newtonian paradigm in physics stood unchallenged until the late 19th and early 20th century. But then came two challenges, one from Einstein's Theories of Relativity, and a second from Quantum Theory. The theories of relativity proposed that experience of events for all observers was not the same everywhere and at all times as Newtonian Physics would have it. Einstein proposed that relative velocities that were a large fraction of the speed of light, and large gravitational fields would destroy the uniformity of the Newtonian world. Subsequent experiments verified the theories, and after 50 years or so of argument the theories of relativity became generally understood and accepted. The situation evolved rather differently for Quantum Theory; first, many physicists contributed to the development of the mathematical models, and those mathematical models were verified just as had happened for the theories of relativity. Quantum Theory has become the most successful physics theory ever for the accuracy with which the mathematical models predict the results of experiments. However, there are still a number of different interpretations of the mathematical models, and to this day there is still no agreed and accepted way of interpreting the mathematics. I will return to this puzzle in Chapter 9.

A few years after the publication of Descartes' treatise, John Locke proposed that separating belief from the senses was utterly preposterous and that Descartes was completely mistaken in his views. Since then, philosophical arguments have raged between the different perspectives, the *rationalist* perspective founded on Descartes' writing and the *empiricist* founded on Locke's view. Subsequently the development of various strands of *idealism*, the idea that the only things that really exist are minds or mental states or both, by Berkeley and others, added to the ongoing philosophical discussions. For those interested, a detailed account can be found in Deely 2001 of these and subsequent developments.

Although the rationalist philosophy is the default in the WEIRD world, challenges to this understanding from other approaches go back to the Ancient Greeks. As I described in Chapter 1, the notion of the flowing world proposed by Heraclitus precedes the static world of Parmenides. Plato's Allegory of the Cave (Plato 1987) is a story of people bound in a cave, only ever experiencing shadows and reflections of the world outside. He described their world as constructed from their experience, restricted to interpreting the shadows and reflections they experienced. When one of their group was taken from his place and experienced the wider world, given time to learn and then returned to tell of his experience, he was not believed by those remaining. The group remaining in the cave rejected the new ideas because it was too far from their experience. Although the story is told by Socrates as an allegory standing for the effect of education, in this sketch Plato is clearly taking a view that the cave inhabitants construct their view of their world and have difficulty in adjusting that view when presented with a different view of which they have no experience. I infer that Plato is saying that "the world as it is" is defined by our experience. But that our experience is limited and we need to allow for the experience of others (other perspectives) as well as our own – for example, the experience of the cave-dweller who was allowed out. So that rather than seeing the world as it is, the implication is that we must learn from our experience, in particular from the experience of others.

7.4 Systems Thinking

Whilst the WEIRD approach had great success in the relatively simple situations found in the material world of physics, it was the desire to be able to model more complex situations, in the social world, that drove the development of Systems Thinking. Modelling interrelationships is absolutely necessary to understand complex societal and ecological situations. Even some relatively simple situations in which interrelationships were fundamental, such as modelling a control system, stood outside the familiar approaches. But, when Systems Thinking began to be developed, like all other scientific disciplines, it was firmly based in the WEIRD world, in the standard Western scientific paradigm (e.g. Ashby 1956, Weiner 1961, Beer 1959). The first definitions of a system in the Western world were (von Bertallanfy 1968) in terms of parts with dynamic interactions between those parts using the language of mathematics; the parts being seen as objects. But whilst the formal definition of system was in terms of an object it was recognised that it was the relationships between the parts, or factors in a situation, which Systems Thinking was reaching towards and attempting to model. At this time, therefore, researchers and authors were grappling with a situation in which their understanding and motivation was to maintain the scientific status quo, developing '*objective*' models but the necessity was to move from the object-based approach to encompass relationships and dynamics. This resulted in various formulations, those quoted above staying within the extant paradigm, but other authors, for example Andras Angyal (1965) and Gregory Bateson (1972), wrote from a perspective that seems to me to be rather different and not coming from an object-based approach.

Perhaps the first systems approach to become widely known was *System Dynamics* through the work of Jay Forrester and others at the Massachusetts Institute of Technology (MIT) (Forrester 1971, Meadows et al 1972, Roberts et al 1983, Morecroft, 2007). It is significant that the understanding produced by the models developed by the MIT group and published in Meadows et al (1972), became widely known, but they were soon to be much derided because the models forecast huge problems for mankind if we did not modify our behaviour to the natural world. However, the predictions made have stood the test of time and the work was considered recently, by a British parliamentary committee for climate change, the single best work of forecasting ever (Jackson & Webster 2016).

The System Dynamics approach illustrated with a simple example in Figure 7.1 starts from constructing what are known as *causal loop diagrams* of a situation. Causal loop diagrams consider the relationships between objects in a situation. The upper part of the Figure 7.1 shows a simple example of such a diagram relating word of mouth recommendations to sales, and production. The information gathered in constructing the causal loop diagram is then transferred to a systemic diagram of processes, stocks and flows, as illustrated in the lower half of Figure 7.1. But in this transition there is, as usual in a WEIRD thinking world, little or no consideration of perspective. In order to move from the object base to the systemic base the first question to be asked is 'what is the purpose of this system?'. As discussed in Chapter 3 different possible purposes can result in very different systemic models. A consideration of perspective is fundamental to process-based modelling and therein lies a hidden weakness in the System Dynamics approach. The second diagram once constructed is then used to construct computer software that is used to calculate and graph the dynamic behaviour of the situation under scrutiny. This technique proved to be a very powerful technique for illustrating the dynamic behaviour of complex situations. It was the first recognition that to model complex situations a network of interconnected processes enabled exploration and better understanding of their behaviour.

Peter Checkland (1981) came up against the weakness of the systemic approaches based within an object frame in exploring the use of Systems Thinking in seeking to understand and manage organisational problems. He found that standard techniques failed and as a result developed the approach which he called Soft Systems Methodology (SSM) and which was reviewed in Chapter 6. This approach takes the key step into the teleological process world by setting a *Root Definition* as the starting point for a systemic process construction. The *Root Definition* in this approach is exactly that which I have defined as purpose in Chapter 3. Soft Systems Methodology explicitly takes into account different perspectives through the mechanism of encouraging the various participants in the situation being examined to define their own root definition and subsequent model. If no agreement can be reached on the Root Definition, the SSM process thereafter encourages the participants to seek agreement on possible action to be taken, without necessarily moving from their own modelling of the situation. Hence the purpose of SSM is to move from a position where participants in a situation have different perspectives on that situation, to one where problem-solving action is agreed, despite those multiple perspectives.

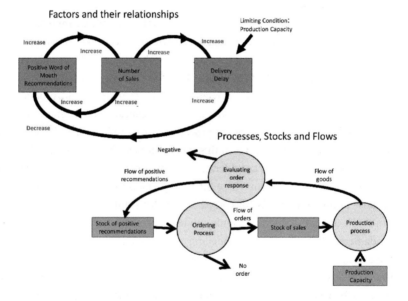

Figure 7.1: The development from the object base to the systemic base

The use of non-teleological terminology in the descriptions of SSM, and the withdrawal of the recommendation of Beer's work as a good starting point for the development of systems models (made in Checkland 1981 but not in subsequent writing), seem to me to result from a perceived need to explain the methodology within the WEIRD paradigm to aid the spread of its use in our current world. The increased use is certainly a positive phenomenon but the loss of awareness of its relationships to other aspects of systemic process thinking seems to me to be detrimental.

Stafford Beer spent time in India during the Second World War as an intelligence officer. During that time he made a study of Eastern Philosophy to add to his previous study of Western Philosophy. This experience gave him a perspective unrestrained by WEIRD thinking and set him on a unique path over subsequent years. The transition from an object base to what I now define as the *systemic process base*, synthesising the experience of Western philosophy and science, with knowledge of Eastern philosophy, can be traced through his writings in the development of the *Viable System Model*. This development starts from *Cybernetics and Management* (1959), written from a Western scientific perspective, to *Heart of Enterprise* (1979) which starts with a clear statement of the subjective process nature of systemic process modelling.

In *Cybernetics and Management* Beer formulates the idea that any management system must have a perspective meta (outside, above) to a system that it manages. This was the first step towards the idea of recursion and the understanding that systems models have a fractal structure. In this he builds on Kurt Gödel's Incompleteness Theorem (Feferman et al 1986) that any formal logical system is always incomplete. There are always questions that can be framed within the defined logical system that cannot be answered from within that system. These questions can only be answered from outside the defined system, that is from a meta-system of that original system. This conclusion led Beer to the proposal that management must always take place from outside, meta to the system being managed (see Chapter 4).

Godel's incompleteness theorem has fascinated many writers, notably Douglas Hofstadter (1979). It is formulated and proven within the paradigm of analytic thinking. But from a systemic process perspective the Incompleteness Theorem can be interpreted rather simply as: *for every system there exists a meta-system*. This aligns with the modelling that I have developed in Chapters 4, 5 and 6. Whatever mindset is present in a brain at a point in time, open learning will build from that as new experiences are synthesised with old, particularly where it is necessary in a changing environment. It is always possible to extend understanding through open learning. This interpretation of Gödel's Incompleteness Theorem is the key idea which led Beer to the fractal (recursive) structure that is a central idea in systemic process modelling. By the time he wrote *The Heart of Enterprise* twenty years later, Beer's shift to systemic process thinking is complete, and he introduces the aphorism "the purpose of a system is what it does". He realised that, in developing a model of a system in focus, not only does an observer's perspective define the starting purpose and modelling process, but just as importantly an observer should not simply accept an extant purpose but examine the situation for themselves to understand what the system does from *their* perspective.

These three approaches to the problem of formulating systemic models in a complex situation, System Dynamics, Soft Systems Methodology, and Viable System Modelling are now three well-established approaches with extensive literatures of use. Their development and use illustrates the move from the use of the WEIRD paradigm underlying Systems Thinking, that is from *systemic object thinking;* to the understanding that the most powerful modelling of an interconnected world is obtained through a *systemic process approach.* However, whilst the first of these three systems approaches has reached public knowledge, and is much used, neither of the last two is as well known. Beer's approach, the most powerful, is the least known but could make an enormous difference to the world in which we live. The model of learning developed in the previous chapter is one contribution; in the next two chapters I examine this approach as applied to the two very different areas of interest which have sustained my research, firstly governmental systems (Chapter 8), and secondly Quantum Mechanics (Chapter 9). These two chapters are but tentative steps in the use of this modelling approach, but it seems to me the promise of improved understanding is very great indeed.

In summary, in my exploration of Systems Thinking I have concluded that, as I noted in Chapter 3, in the early days of Systems Thinking systems were defined as close as possible to the object frame. But this is difficult because in this framing the object has no connections to its environment and results in important connections being missed. I introduced several examples in Chapter 3, where building models based on purposeful systems enabled a differentiation to be made, when in typical Western thinking no differences are easily detected. For me the most startling example is the way in which commercial companies are defined as having the primary purpose of making a profit, which has led to potential self-destruction due to the pollution produced by our way of life which has been overlooked because of our way of thinking (Carson 1962). In contrast these polluting flows cannot be overlooked when thinking in terms of systems. In contrast to the WEIRD way of thinking, a *system* is defined by its connections to its environment, together with how it changes its environment, and is changed by its environment (as discussed in Chapter 3).

7.5 The Shift to Process Thinking

The focus on process as the centre of the systemic definition *input→process→output,* and the systemic process models being interconnected networks of processes, brings the concept of process to the fore. The most important contributor to an understanding of process and its associated philosophy was Alfred North Whitehead (Whitehead 1978). The first publication of *Process and Reality* in 1929 preceded the development of Systems Thinking described above. Whitehead's formulation is rather difficult to follow, but it does relate directly to the systemic process formulation I have developed here. Figure 7.2 illustrates Whitehead's formulation of the process of perception to action in the diagrammatic language I have used. Whitehead's central idea is an 'actual occasion', that is a situation point in time where an animal perceives the need to respond.

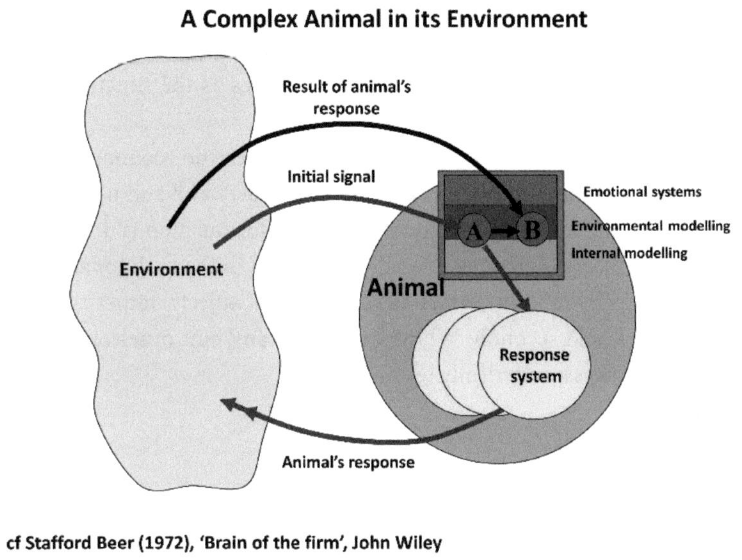

A Complex Animal in its Environment

cf Stafford Beer (1972), 'Brain of the firm', John Wiley

Figure 7.2 Whitehead's Process

In Whitehead's formulation an '*actual occasion*' comes into being by drawing on, in his language 'prehending', past actual occasions in a process of 'concrescence'; the many past actual occasions coming together to form the one present actual occasion. In Whitehead's nomenclature the 'this' and 'that' of '*subject*' and '*object*' now refer to the current actual occasion and the past actual occasions. When the subject actual occasion B prehends an object actual occasion A, it imposes its perspective. As a result this process draws from A only the aligned elements of data, and eliminates those elements of A which are not aligned to B's perspective. Whitehead refers to this as the carrying over of feelings from A to B. The use of the word '*feeling*' extends the usual usage but carries the same meaning e.g. a feeling of anger might be carried forward from one occasion to another.

Whitehead's terms, *actual occasion, subject, object, feeling,* are evidence of an object-based approach which for me makes the language and formulation of the ideas at odds with the underlying ideas he wishes to convey. Donald W. Sherburne (1966) describes Whitehead's approach as an 'atomistic system' referring back to the ancient Greek Philosopher Democritus, the originator of the idea of a fundamental particle of matter which is of course in concept an object. Not surprisingly, attempting to describe processes in terms of objects results in a hugely complicated set of explanatory models. However, the systemic approach developed by Beer, which is the foundation for the exploration that I have undertaken, conceptually aligns with the processes that the modelling is seeking to represent. The necessity for a modelling language/representation to have the same symmetries as the situation being modelled is further discussed in Chapter 9.

Figure 7.1 illustrates how the two approaches do connect. Beer's formulation is simpler, and the models easier to understand and use because the formulation aligns with the ideas. Whitehead's formulation is difficult for the two reasons that it is an object-based formulation in the traditional way of attempting to understand processes as secondary to objects, rather than the other way round, and secondly Whitehead avoids any hint of teleology even though clearly processes are firmly teleological.

7.6 Constructing a world

In the systemic process approach the input and output flows interconnect processes and define the relationships between them. As a result, in using this modelling approach the resulting systemic process models form a complex interconnected network. Moreover, the flows between two processes often operate in both directions, forming a feedback loop. Further, in general, many feedback loops can be traced through the interconnections between multiple processes, so that feedback loops are ubiquitous in such an interconnected network. A society is such an interconnected network; societies are made up of the relationships between people, between people and organisations, and between organisations which are themselves made up of the relationships between people. But of course, human societies exist within, and are part of, the larger interconnected systemic network of all life on planet earth.

A brain and nervous system is also such an interconnected network, in which mental models are formed by the signal flows through the structure of this network. The structure and flows define both the way we think, and what we think. These models are the means by which we recognise particular situations and produce rules for the actions we take. Since the mental models themselves are processes, for simplicity any modelling approach and underpinning philosophy should be one that has a systemic process base.

An internal model is developed so that an animal can predict the state and configuration of the environment in the future, given the state and configuration at a given moment, from the viewpoint of that animal at that moment in its life. The model enables the animal to survive, fulfilling its purpose in its econiche. In life there are many surprises, so a model will not *guarantee* survival but as long as it enhances the probability of survival a particular model will persist in the animal's repertoire. The purpose of such a process model is to predict how the environment will flow, not what things are. No animal comes into its world with a mind which is a blank slate; rather it arrives with the models that evolution has given it. No individual living form starts from nothing; all come into the world with a starting set of inbuilt models but, as soon as possible, they must start the process of learning. A calf born on the savannah can be up and running within minutes of its birth, and then must start learning survival strategies as an individual in the herd. A human child starts its individual learning in the womb, then after birth, increasingly exposed to a situation and a culture into which it has been born, learns in small steps, adapting to and with the world around it. It inherits human form but adapts, in time propagating both its genes and the culture. Jean Piaget describes the early stages of this process in *The Construction of Reality in the Young Child*. His descriptions of a child's progress reflect strongly the modelling I described in the last chapter.

But from this evolutionary start there is still an immense amount to learn. For any animal, the perceived world out there is complex beyond the capacity of its mind/nervous system, and in learning to cope with its world it is faced immediately with the same three difficulties that produced the understandings of Systems Thinking. These are:

1. to begin to deal with the world out there we must make sense of the jumble of signals being received by our sensory organs. To make sense of those sensory inputs we must look for constants in the flow of signals so that we may construct boundaries differentiating out sections which carry some coherence and constancy. So, the first criterion to be taken into account is the judgement of where boundaries are in the jumble of sensory signals we receive.

2.   Having selected those boundaries, the world out there which must be appreciated and understood then consists of a mass of interconnected parts. The realisation of this is the second criterion.

3.   The last criterion, and the most important, is the understanding that perspective plays a fundamental role in how and what we perceive. And it is an examination of this third criterion that enables a way forward in understanding why WEIRD thinking falls into the trap of not discriminating differences in the way I explored and illustrated in Chapter 3.

My conclusion from these considerations and the previous chapters is that animals construct their world (*constructivism*) through the mechanisms outlined in Chapter 6. Observing my cat, I understand how from observation he comes to model my behaviour, anticipates my actions and is able to manipulate my behaviour. But my cat and I have the disadvantage of not being able to communicate very effectively. I don't always understand what he wants, nor he what I want. Learning takes longer on both sides and is less efficient. It is the ability to communicate in a very sophisticated way that sets *Homo sapiens* apart. Although other animals also learn from the experience of others of their species, our experience includes the ability of others to challenge or reinforce our models in very sophisticated ways. There is always a variety of experience amongst individuals of any species, and ideas do spread if there is communication, as in the world of apes, primates, crows and others even though there is a lesser variety of communication.

As a result of these conclusions, I had the revelation that for the ancient Greeks their gods were real, as real as electrons, protons and atoms are for me today. But what is 'real'? If I think about Greek gods or electrons then I am invoking a model that I have stored in my brain. The difference between real and not real in these examples is that for me the particle models (electrons, protons and atoms) have been introduced as real and subsequently reinforced as real in all my interactions with books, people and experiments. Therefore, I accept that situation and speak, write and act on the basis of that view, reinforcing that view in my interactions with others. On the other hand, over my lifetime I have met the Greek gods as the beliefs of ancients, as interesting stories that don't stand up to my expectation of experimental verification that I wish to see following my training as a Western scientist. Therefore, I speak, write and act on that basis, reinforcing that view in the society I live within. Had I been born in ancient Greece I would have been introduced to the models of the gods and they would have been reinforced by society, just as the models of electrons, protons, and atoms are today.

How do these beliefs come about? The problem is that the world out there is complex beyond our human ability to model it. We can only construct our worlds from our experience, and therefore each of us lives in a world that is unique. Our models are formed both from our experience of our cultural world in the feedback loops between ourselves and other people, and from our experience of our environment. Because in general we can agree on the simpler things, many believe that what we perceive is what there is. But WEIRD thinking extends this modelling process – which enables us to successfully survive in our physical econiche where individuals *do* agree – to explain other more complex phenomena where individuals do not agree. But even if we do agree, about atoms and electrons today, for example, and about gods in ancient Greece, we still need to look very carefully at what we mean by 'real'.

In 1910 Cassirer wrote that that the concept of atom had the "logical function" of relating together the results of various experiments and as a result he called it a "mediating concept" (quoted in d'Espagnat 2003). He considered the concept a useful device, but not that atoms are 'real'. This was remarkable for that time, as physics was still very much based in WEIRD thinking. However, since then Quantum Theory has reinforced Cassirer's view calling into question any belief in the concept of particle as 'real' but certainly supporting that it is a useful "*mediating concept*" in some circumstances. Thus some concepts such as atoms and electrons in our society, or Greek gods in ancient Greece, usefully explain – in the extant culture – observed phenomena, but do not necessarily survive a change of culture.

I must conclude therefore that whilst models are constructed in a feedback loop between an animal and its external environment it is not possible for those constructed models to exhibit the variety of that external environment. Thus, in seeking a supporting philosophy, it is clear that philosophy cannot be the realist philosophy that is at the root of WEIRD thinking. The philosophical approach to understanding understanding I am taken to is the *Pragmatist* philosophy originated by Charles Sanders Peirce. William James states that Pierce's Pragmatist philosophy proposes that our beliefs are "*rules for action*", and to develop a thought's meaning it is sufficient to "*determine what conduct it was fitted to produce; that conduct is for us its sole significance*" (W. James 1907 in Thayer 1982, p.210). In the systemic process approach, mental models are indeed rules for action and it is the resulting conduct that is the sole significance. Whilst this approach was originated by Peirce, it was subsequently developed by William James and John Dewey. James further says that:

> There can be no difference anywhere that doesn't make a
> difference elsewhere – no difference in abstract truth that
> doesn't express itself in a difference in concrete fact and in
> conduct consequent upon that fact, imposed on somebody,
> somehow, somewhere, and somewhen. The whole function of
> philosophy ought to be to find out what definite difference it
> will make to you and me, at definite instances of our life, if this
> world formula or that world formula be the true one.

I interpret what James means by "world formula" as the set of mental models in use at any one time, and "true one" as the viable model in the sense I have defined. Pragmatism as defined by Peirce, James and Dewey is therefore a very practical philosophy, which fits the systemic process approach to understanding, namely that the processes of evolutionary learning and individual learning construct understanding and response to environmental circumstance.

What is important is that the more precise thinking given by a systemic process approach really does make a difference in action. As James argues:

> ...the tangible fact at the root of all our thought-distinctions,
> however subtle, is that there is no one of them so fine as to
> consist in anything but a possible difference of practice. (1907)

But from this systemic perspective there is no "true one" as James has it. Each of us has a unique set of lifetime experiences from which we may learn and, therefore, a unique set of models, and in any situation there are just models which are more or less useful. This then takes me back to the Conant-Ashby theorem which, in essence, states that situations are successfully survived only if understanding is sufficiently good, and therein lies the benefit of being able to communicate sophisticated models between individuals.

Ernst von Glasersfeld (1995) has developed Radical Constructivism, a modern version of this pragmatic approach to learning and understanding which he characterises as follows.:

> Radical constructivism is uninhibitedly instrumentalist. It
> replaces the notion of 'truth' (as true representation of an
> independent reality) with the notion of 'viability' within the
> subject's experiential world. Consequentially it refuses all
> metaphysical commitments and claims to be no more than one
> possible model of thinking about the only world we can come
> to know, the world we construct as living subjects. (1995)

Radical Constructivism aligns with Pragmatism in its instrumentalist approach, but changes the understanding of 'truth' to experiential 'viability'. This use of the word 'viability' matches exactly that of the use I have made in following Beer's use in the development and use of the Viable System Model. A mental model is viable if it stands the test of experience. Hence, I conclude that with the foundation of Pragmatism, originating with Peirce, through the work of Piaget to von Glasersfeld's Radical Constructivism there is a coherent logical foundation for systemic process thinking as I have described, which replaces the reliance on 'realism' in the underpinning of WEIRD thinking..

7.7 The Relationship between System and Object

In Chapter 2 I laid the foundation of the systemic process modelling approach by exploring the idea of a model. The chapter moved to defining a model as a system that produces a particular output from a given input aligning with the *holistic reasoning* of Henrich et al (2010):

> Holistic thought involves an orientation to the context or field as a whole, including attention to relationships between a focal object and the field, and a preference for explaining and predicting events on the basis of such relationships.

In Chapter 3 I explored the idea of 'system' and found that hidden in the mathematical ideal there was great complexity. The same *object*, a tiger, a railway or a company, can be modelled in different systemic ways. We each construct our own world from our own personal experiences. Each of us has, as a result of our own unique set of past experiences, a mindset of mental models from which we construct our own *perspective* on any situation we encounter. This means that the same pattern of signals will be perceived differently by different people; each will perceive and interpret that pattern according to their particular mindset. Each will construct models of the same *object* from their mindset which reflects their *perspective*. An accountant designing hospital systems with the purpose of minimising costs will design those hospital systems rather differently from a doctor designing hospital systems with the purpose of providing the best patient experience. In these two cases the hospital's relationship to its community will not be the same. This is not to suggest that each of these people is unable to embrace the other perspective through learning, but that we are dependent on the mindset we have in the moment. As I described in the last chapter, a mind evolves by building slowly on what is there already.

Figure 7.3: Ahead, a Roman Millarium

In early 2017 I was walking north with my partner on the Via de la Plata on our way to the small village of Aldehuela de Jerte when we were hailed by Rob, a Quantum Physicist retired from Imperial College, walking south. A discussion ensued on the problems of modern Quantum Physics which seemed a little surreal at the time standing by the roadside on a showery day. That conversation set me thinking for the next few days. Then suddenly as I approached a Roman milestone (millarium) I finally understood the relationship between an object view and a process view, and why this has been misconceived by WEIRD thinkers. Ahead I perceived a tall rectangular block of stone, which was that Roman milestone. I realised that I perceived this stone block because it scattered the light falling on it, and some of the reflected light entered my eyes. I discriminated this out from the background scene, attributing a boundary in the perceived pattern. This particular perceived pattern of the milestone was defined by that particular place where I stood and its particular relationship to the milestone. In that moment I realised that I *should* model the milestone that I perceived as a *system*. There was an input, the flow of ambient light onto the milestone, and an output from the milestone, the flow of reflected light, some of which was received by my eyes. It was the disturbance to the flow of ambient light, the processes of reflection and absorption that enabled me to perceive the milestone.

My Western mind had always assumed that an *object* is perceived. The question was, therefore, how does a Western mind get from the single systemic observation as described above, to the assumption of object? In making that observation, viewing the milestone ahead, I had discriminated out a boundary in my visual field and focused my attention. When I changed my position to a different view and maintained my attention, the scene and my relationship to the milestone changed. That second *observation* of the milestone, whilst from a different *perspective*, has commonalities with the first observation.

The important leap of understanding was that if I was viewing that system for the first time, my brain would then automatically synthesise the perceptions from these two observations using *Open Learning* as described in the last chapter. I would be developing a schema-system from the synthesis of the commonalities of my two different observations. I would go on to take more observations, view the milestone from many different perspectives, all the time synthesising those observations with the previous observations, enhancing my schema-system model. As Piaget (1954) describes, this normally happens naturally, early in life, when an infant starts to move about, changing its perspective on the world it finds itself in. And, of course, I have no reason to suppose the same processes do not happen in other animals capable of learning in this way.

But when something becomes very familiar from those many observations, in the end I forget that each time I view that thing, I do it from a particular position. I make a particular *observation* from a particular *perspective* which has a particular relationship to whatever it is that I am observing. But it is that single step, the single observation, which is fundamental – my 'object' model is built from many such single steps. Once my brain has built that schema-system model, it recognises the upright cylinder, or any other object, very quickly from the pattern of scattered light reaching my eyes from any observational perspective.

The use of the word *perspective* here denotes viewing from a particular spatial point: the geometric interpretation of the word perspective. For each position I stand in, for each perspective I take, the pattern of reflected light reaching my eyes is different, and only a small part of the reflected light emanating from the milestone. It is my brain which then over time, using open learning, synthesises these different patterns to form the schema-system model. The schema-system model and those many relationships are telescoped into no relationship at all in the abstract. But the model in my head is still a schema-system model ready for use when a Roman milestone appears in my visual field. The processes of synthesis of different perspectives into an object model honed by evolution works well in perceiving and reacting to events in the natural world, but when we come to complex situations, for example understanding a functioning organisation, with variety beyond our capacity to span, we are misled by our assumption of object. We are each limited and encased in our own limited 'bell jar' of experience (Plath 1966). Provided we are open to learning, those limits expand with our learning as Perry describes (see Chapter 6).

Underlying my systemic process approach is that 'out there' is understood as a network of interacting systems that I perceive, but models are developed to the point that systems can be recognised from multiple perspectives and idealised as objects. Thus, there is always an interplay between system and object. Systemic process models are the fundamental building blocks from which object models are constructed. That models are constructed either through lifetime learning or species learning becomes the central tenet of understanding the development of a brain and nervous system. Thus Daheane in the quotation above is incorrect in his answer to the question he poses. We do learn to see objects. This works well in coping with the world around us, but we then carry this learning into situations in which it does not apply.

In any situation wherein I am limited by my brain's capacity, the relationship between an object and a system is analogous to the relationship between zero and the rest of the world of numbers. For every assumed *object* there is, if not an infinity, a large number of possible *systems*, depending upon the perspective taken. Therefore, framing the world in terms of objects is going to result in a rather simplistic world view compared to one built on being aware that those possibilities exist. As I have discussed, there are vital differences between systems constructed with different purposes in mind, and as Beer (1966) notes, discussions between politicians with different perspectives cannot reach agreement. Evolution has equipped *Homo sapiens* with a more sophisticated brain and nervous system than other mammals, but it evolved in a context of small hunter gatherer tribes. Whilst there were differences in aspects of culture across the world, those differences were seldom encountered, and lifestyles had much in common. Technology has massively increased our ability to communicate across both time and space but also the variety of life experiences. A city banker has little life experience in common with a migrant agricultural worker. We assume that a communication once sent will be understood as we intended. But that is hardly ever the case, as I noted in Chapter 2, where I introduced the enhanced communication model (reproduced at Figure 7.4) to illustrate just this understanding. When we discuss communication we conflate this with transmission and have no way of checking what was received. As I noted in Chapter 2, a centralised government or management can broadcast a message but cannot close the feedback loop with each and every recipient to ensure a received understanding.

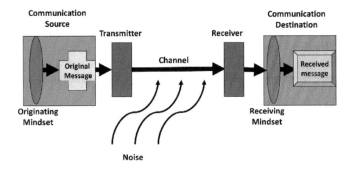

Figure 7.4 The Enhanced communication model

7.8   The Essence of Scientific Method

I have reached the point where it is now possible to understand why I am able to turn Nagel's (1961) argument on its head as I asserted in Chapter 1. Nagel claimed that four philosophical approaches underpin science; two of these are based in the object world and two in the teleological process world. He asserted that the second two could be derived from the first, but for me the derivation is flawed. I now have turned the argument on its head and shown how the Western scientific and philosophical approach and the foundation of Parmenides can be framed within the foundation of the approach of Heraclitus. Objects, the basis of WEIRD thinking, the Western scientific approach, are constructed from observations of systems, and built as a simplification of the teleological process world. In hindsight it is not unexpected that the basic and 'natural' way of thinking is that practised globally throughout history and most of the world. Also, as I said in Chapter 1, it might be expected that the Western way of thinking arises as a sophistication which enables flexibility, but if extended beyond its use for survival in the natural world, pays a price by losing precision and losing sight of the dynamic interconnectedness of the world around us. It is not surprising that, if I am trained into flexibility and simplicity, it becomes very difficult to get a grip on complex interconnected problems.

Science is a process, the process of learning about the environment that we find ourselves in, about ourselves, and the relationship between those two.

Both Karl Popper (1972) and Thomas Kuhn (1996) describe the importance of problem-solving in the creation of new ideas and new models in the advancement of science. Problems crop up especially when we are disappointed in our expectations, or when our theories involve us in difficulties, in contradictions; and these may arise within a theory, or between two different theories, or as the result of a clash between our theories and our observations. Moreover, it is only through a problem that we become conscious of holding a theory. It is the problem which challenges us to learn; to advance our knowledge; to experiment; and to observe. (Popper 1972)

Problem-solving starts with an individual becoming intrigued by a particular problem and worrying at that problem, and very often discussing it with others. But individual learning is the starting point for Scientific Method. New models are proposed and explored which enable the prediction of how our environment might change depending on the actions we take. The models enable the engineering of substances, machines and constructions; and they have enabled increases in energy usage and food production. The guiding principle is that of Pragmatism, namely that thought distinctions are meaningless unless they lead to a difference in practice. A scientist is a professional sceptic concerned with testing models as much as proposing them. Karl Popper formalised the testing process: once a new model is proposed it is the scientist's obligation to devise ways of testing the validity of that model. If the model survives the testing then there is an obligation to broadcast the model for others to test. At any stage the model can be rejected.

Figure 7.5 is based on Figure 6.7 (the Learning Cycle) adapted to reflect the current context. I might write that it illustrates a scientist reflecting on a particular problem and proposing and testing a model, perhaps many times round the cycle. But there are two important caveats that need to be stated. First, in the spirit of the subjective basis of the foundation I have constructed, I should write that I carry out my investigation into that problem to the point where I believe to my satisfaction that I have achieved a new understanding. I have formulated a new model, tested that model as far as I can and I am satisfied that the model has remained viable in withstanding my tests. That new understanding, that new model, I must then communicate to others. Others must then test that understanding, destroy it or be convinced that the model is viable and take on that new understanding.

Eventually a model either converges to a viable model of the phenomena concerned or is rejected as a result of the intersubjective activity and communication. Of course, there are always other factors at play. Just as in the allegory of Plato's cave it takes time for new models to be accepted. Thomas

Kuhn explored this process, characterising the transition from one explanatory set of models to a different set with changed interpretation of a body of evidence, as a change of paradigm. And, inevitably, with an ever-increasing body of human knowledge, subcultures exist within the scientific milieu. A paradigm change takes time to propagate across subcultures. The second caveat is that the model in consideration must be a process systemic model as described in Chapter 3. Only if it is can it then be synthesised with other such models of the same phenomena from other experiments.

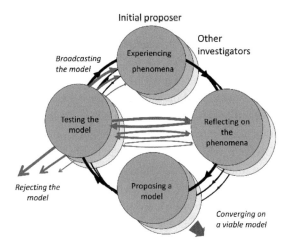

Figure 7.5 Scientific Method as an Ongoing Process

There is no need of the infinite regress of modelling modellers: this is a problem only for WEIRD thinkers. To reiterate, this description is deliberately framed in the first person since that is my perspective. From my testing I believe this is a viable model. It is now up to others to test the model. What results from this process is gradual intersubjective agreement on the new understanding or an acceptance that this exploration was a dead end; it is unlikely that nothing new will result. Scientific Method is a group learning process and therefore is the basis of group acceptance of what we might designate 'objectivity'.

7.9   Whitehead's Tests

Whitehead (1978) proposed that any speculative philosophy should satisfy four criteria: it should be coherent, logical, applicable and adequate. The approach I propose is the result of more than seventy years of learning, seeking explanations for the patterns I have experienced. I propose, building on a foundation proposed by Beer, an approach to modelling complex interconnected situations which can be applied generally to human experience. This comes from the set of models from the many thought experiments I have carried out. I have sought examples where the model gives explanation. The models have been for me confirmed by my testing and this testing has often improved and honed the approach. They have also, to my mind, brought increased coherence to areas of understanding, so I now wish to communicate those models to others. It is for others then to test the model in the same way as I have done for verification. For me the approach is coherent, and logical, and certainly applicable to those areas I have researched. Time will tell whether it is adequate as a speculative philosophy.

In the West, we act as if we believe that an exponentially expanding population, each of us expecting living standards to improve, can be sustained, even when it is based on extracting non-renewable resources on a finite planet. Since this belief is a clear piece of nonsense, if we are to sustain a human population without war, disease and starvation, the way we act must change. Therefore, our beliefs and our models must change. A new way of thinking, a new paradigm, is essential for sustaining a human world. Thinking in terms of process or systems – which is the focus of this book – enables modelling which brings out the interconnections that are missing from WEIRD thinking. In Chapter 4 I began to explore the way in which a systems approach can enable understanding of how, in a world of movement and change, there can be stasis, and so to develop my central argument. This central argument is that the Western world view, the WEIRD approach, is not as useful as a systemic approach when dealing with the many complex situations in the world in which we now find ourselves. In Chapters 5 and 6 I hope that I was able to show at least the beginnings of how this approach can illuminate the complexity of a living learning organism, not the least of which is each one of us, examples of the species *Homo sapiens*.

I was not surprised to find, in my teaching, that many students were natural systems thinkers, given that a brain is a systemic processor. Some, however, found the ideas very difficult to grasp, and a few rejected the ideas altogether. When I examine the work of those who have contributed to the development of Systems Thinking I find writers on both sides of the object-process divide. In my following two chapters I examine the application of these ideas in two areas which have been of great interest to me through my life.

Part 2: Applying the Approach

8. Modelling Governing

Our social systems are far more complex and harder to understand than our technological systems. Why, then, do we not use the same approach of making models of social systems and conducting laboratory experiments on these models before we try new laws and government programmes in real life? The answer is often stated that our knowledge of social systems is insufficient for constructing useful models. But what justification can there be for the apparent assumption that we do not know enough to construct models but believe we do know enough to directly design new systems by passing laws and starting new social programmes? I am suggesting that now we do know enough to make useful models of social systems. Conversely, we do not know enough to design the most effective social systems directly without first going through a model building experimental phase. But I am confident, and substantial supporting evidence is beginning to accumulate, that the proper use of models of social systems can lead us to far better systems, laws, and programmes.

(Jay W. Forrester 1971)

The key issue here is how to reorganise the global economy along social federalist lines so as to allow the emergence of new forms of fiscal, social, and environmental solidarity, with the ultimate goal of substituting true global governance for the treaties that today mandate free movement of goods and capital.

(Thomas Piketty 2020)

8.1 Introduction

My interest in government was sparked by the head teacher at the last secondary school I attended, but at that time my interest was overshadowed by my enthusiasm for mathematics and science. However, the experience of living and working in the United States from 1968 to 1970 reignited that interest, so that on my return to the UK I began a programme of reading to better understand government and its functioning. Apart from the work of Jay Forrester (1971) and the MIT group that was discussed in the last chapter, it became clear from my research that science had had little impact on models of governing and global evolution. This remains true to this day but there is now increasing interest in applying Systems Thinking to the endeavour of modelling government (e.g. Ison and Straw 2020).

As a result of my experience of the United States I joined the UK Labour Party in 1970 and in 1974 I became a party activist, and so began my initiation and apprenticeship into the complex world of management, government and democracy. In the first instance those experiences raised many questions for me, but as a result of an article in a computing magazine describing project Cybersyn, undertaken by Stafford Beer in Chile 1970-1972, I read my way through his writing. The Cybersyn project was described by Beer in his book *Brain of the Firm* (1972/1981) and has been subsequently described in *Cybernetic Revolutionaries* (Medina 2012). In 1983 I sought to meet Stafford Beer because, in his writings, I believed that he had created a scientific path to understanding the questions that my previous reading and practical experience had raised. That initial meeting with Beer and the many subsequent others until his death in 2002 resulted in the change of direction to my life from exploring the ideas of mathematics and physics to exploring Systems Thinking and its applications.

A society is a dynamic entity that exists in a dynamic world and should be conceptualised as a dynamic system; that is, it should be thought of as a purposeful input→process→output structure consisting of subsystems defined as I have discussed in previous chapters. Additionally, modelling governing is exactly to model a control system with the purpose of holding aspects of the dynamic society steady, as also discussed in previous chapters. Governing, therefore, is an application of the science of cybernetics, just as Plato (2006) more than 2,300 years ago and Ampère (1834) nearly 300 years ago both understood. Any governing system must be structured in accordance with the laws of cybernetics; in particular that is Ashby's Law of Requisite Variety, the Conant-Ashby theorem, and the sub-optimisation theorem all discussed in Chapter 4. We would not now neglect the laws of gravity and aerodynamics in designing an airliner, but we do the equivalent, forget the contributions of Plato and Ampère, and neglect the laws of cybernetics, when thinking about the design of a governing system. Not surprisingly our governing systems don't get off the ground!

As with all WEIRD thinking, traditionally the starting point when thinking of government, is to think in terms of objects, hence *nation, country, government, parliament,* etc. – words all describing objects not processes and omitting the fact that all exist within the dynamic ecosystem of planet earth. But government is about leading and managing a society, it is a regulatory system which must sustain a group of people and their environment. The world is a dynamic place, the environment is an evolving dynamic system; the collection of people that are present in the geographic area that is a nation are an evolving dynamic system. The culture, attitudes and values of these people are emergent properties of that system and change over time. The actions of the people can change the ecosystem for the better and for the worse. A society must maintain a productive relationship with its environment if it is to survive. The government is also an evolving dynamic system itself, and the culture, attitudes and values of those people who form the government are emergent properties of that system and change over time. One of the purposes of democracy is to keep the evolution of the governmental system in step with the evolution of the system that is the nation as a whole. A government must maintain a productive relationship with its people if it is to keep their trust. If it parts company with its people the historical evidence is that even the harshest and most violent methods cannot maintain a government in power for very long. Governments of this nature inevitably fail but usually not before many people have perished. A representative parliament therefore should be an evolving system that is capable of reflecting the evolution of its people. To that end we could ask such questions as 'why have all elections on the same date?', 'Why have fixed term parliaments with all members potentially changing at the same time?'. There are many other possibilities that would provide steady evolutionary change. Are elections the best method of choosing a representative? The ancient Athenians used sortition, a lottery system, to choose their government. A lottery system takes away one of the least desirable aspects of our present system: that representatives are chosen on many occasions from those who seek power for personal reasons, and once they are elected seek change in order to preserve their own continuing position holding on to the power they have achieved.

Societies are of course extremely complex in their functioning, much more than any airliner. My aim for this chapter, therefore, is just to outline an approach to modelling governing from a systemic process perspective that is rooted in the science of cybernetics. I begin by setting out how a collection of people may be defined as a system in the same way that any living entity may be considered in an ecological system. From there I then first consider the implications for the way in which the environment might be managed, following the considerations in Chapters 5 and 6. I next explore the guiding implications arising from the Law of Requisite Variety for a constitution. Further, the Conant-Ashby Theorem tells us that to be effective a control system must contain the best possible model of the system under control.

Because of the complexity of a community, only by engaging all minds in that community can an effective model be brought to bear on the problem of controlling the community. This I take to be the primary purpose of *democracy* and why democracy is thought to be the best possible form of government. But of course it is *through* the governing system that all minds are brought to bear, and there are a multiplicity of ways in which the governing system can be structured to achieve this aim: the Conant-Ashby Theorem and Beer's VSM suggest the guidelines. From my studies and engagement it seems to me that in the great majority of cases the systems of national government currently operating are failing. I therefore explore the consequences of using the principles developed in the previous chapters to design guidelines for a *process systemic democracy*. Lastly I explore the implications of the sub-optimisation theorem, how can we avoid sub-systems seeking to optimise their wellbeing, wealth or any other variable at the expense of other sub-systems.

8.2 Defining the system in focus

Nations are a relatively modern idea (Hobsbawm 1990) and assumed to be, if not the sole focus when considering government, at least the principal focus. In Chapter 4 section 4 I explored the question of the understanding of the systemic individual and concluded that an individual system is the system in focus at any one time. I therefore propose to consider government in terms of *community*. A community can be a band, a tribe, a chiefdom, a state (Diamond 2013), but further than that also a hamlet, a village, a town and many other types of group. I begin with communities that live together in a geographic area; later I will consider the role of communities brought together by their common skills, and other organisations operating within a geographic community or across many geographic communities.

Communities have a very much longer history than states or countries and certainly predate agriculture and the idea of fixed settlements. Early communities were small groups of people but in the modern world they can be of any size, even, a country, a region, or the whole community of *Homo sapiens*. Further, in our modern world, communities come together not only in terms of communal living, i.e living together in a geographical area, but also coming together through common interests and common expertise. The problems to consider in governing a community include: what is the purpose of the community? How must communities come together to build larger structures? And how might the various communities work together?

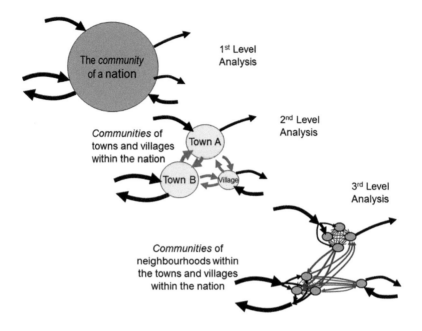

Figure 8.1: Illustrating the fractal nature of the word 'community'.

Just as animals and ecological systems sometimes fail to adapt, communities do too, and sometimes collapse. Malcolm Levitt (2019) concludes from the historical evidence that:

> Explanations of collapse in terms of competing mono causal
> factors are found inferior to those incorporating dynamic
> interactive systems.

This brief extract supports the path I am taking, that a systemic approach to understanding how a community might maintain viability is the one most likely to provide explanation. Levitt also writes that:

> ...collapse should be explained as failure to fulfil the ancient
> state's core functions, assurance of food supplies, defence
> against external attack, maintenance of internal peace,
> imposition of its will throughout its territory, enforcement of
> state wide laws, and promotion of an ideology to legitimise the
> political and social status quo.

Here Levitt sets out the purposes that the government of a community must address. In summary this is exactly the maintenance of the relationship with the external world and the maintenance of the internal relationships as would be expected from the preceding chapters. But he does not mention here the role of the natural environment which has played a role in societal collapse (Diamond 2005)

The problem for management of a human organisation is to constrain the variety of actions in the behaviour of those people involved within the organisation to those actions which fulfil the purpose of the organisation. But if we now focus on a community and its government, for a community of people co-living there is no immediately obvious overall purpose to constrain actions to in quite the same way as for example the restaurant considered in Chapter 4. From the preceding chapters I conclude that any human organisation has the same needs as an animal to survive and adapt to a changing environment. But any human organisation has the added problem of maintaining its internal structure and the necessary relations between all the people (and organisations including sub-communities) involved in that community. People are free agents in a way in which the sub-systems of an animal or ecological system are not. Evolution has reduced the variety of states of the subsystems of an animal or an ecological system to very few states leaving little to be done by the managing brain and nervous system even if there is one. There is a much more complex set of interrelations to be maintained between the subsystems of organisations and people in a community. Firstly, I will consider the maintenance of the external relationships, and then subsequently the maintenance of internal relationships.

8.3 Community and Purpose

For any animal the first priority is survival; survival for an animal means survival in its specific econiche. That econiche is contained in an ecosystem which in turn is shaped by the local geology, the shape of the land, the local weather, and the co-living species. The species *Homo sapiens* evolved from its ancestors as a social animal. New members of ancient and modern hunter-gather groups are born into an ecosystem and, as children, learn how to live by watching and listening to members of their group. At some point in the history of *Homo sapiens* and its ancestors the skills of manufacturing and using tools were discovered and developed. The development of these skills increased the variety of ways in which group members could relate to one another, and therefore increased the potential states of the group. The important point to note here is that in essence skill development and maintenance lies at the heart of trade and trading relations between communities. In the first instance the skills were those which exploit natural resources: the development of the ability to manufacture hunting tools, pottery, clothing, and the ability to cultivate crops. But, of course, today the variety of possible trading relationships is enormous.

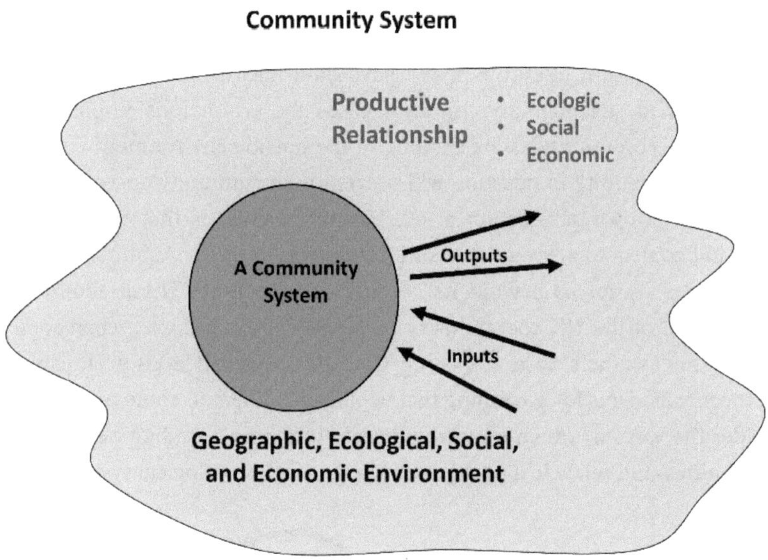

Community System

Figure 8.2: A community in its environment

A community seeks to survive, but in what way is the community going to relate to the environment in which it resides? I write these words sitting in a small apartment in a town which decided in the 1970s to take advantage of its immediate coastal environment and determined that it would become a seaside holiday resort. It embarked on a massive programme of constructing many small apartments, and also a marine lake for families and others not wishing to brave swimming in the Atlantic Ocean. In the winter it has a population of 1,700, in the summer 25,000. By doing this it became prosperous in the current world, but of course the question now is: can it adapt to the changing world of increasing sea levels and fiercer storms? In this case both the individual community and the species of seaside holiday resorts is challenged.

My home town was a prosperous industrial town of the industrial revolution that has been long in decline. In 2017 two community enterprises came together to attempt to rescue the town's semi-derelict community college to teach the theory and practice that will be necessary to counter climate change. It is an ambitious project to start to build a new green economy, and an attempt to answer the question of how my home community is to relate to the outside world in a new positive way. The question of how a community should relate to its outside world is a question that should be asked and answered for all communities of every type and size. What is this country, region, city, district, town, etc for? Or what does it wish to be? These are not questions that I have often heard posed, but they are fundamental to sustainability. In most cases the restrictions imposed by ownership of community assets residing in the outside environment and not inside the community in question will prevent the community answering the question to its own satisfaction at all. History teaches us that without the ability and control to answer the question for itself, then a community will be overwhelmed by forces beyond its control at some time. Think about the closing down of the UK coal industry in the 1980s for example, when some communities lost their reason for being. Communities at all levels in large part lack the mechanisms for governing that would allow them to come together to consider the ways in which they would seek to fit into a trading network of communities, and relate in a positive way to the earth's ecological systems.

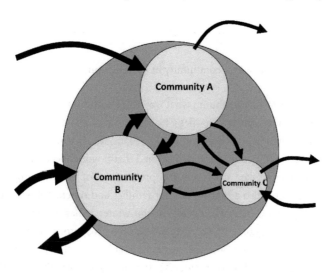

Figure 8.3: Trading communities

In practice what does this mean? How far could a community actually be self-sufficient? In our modern world most communities are not self-sufficient but need to contribute to the outside world in order to take in goods and services which they are unable to supply for themselves. How will each community relate to other communities in the outside world – what trading relationships should it establish and what should it do for itself? Many communities struggle unsuccessfully against the almost completely unregulated feedback loop of capital creating more capital which has led to over-centralisation and exploitation (Piketty 2020). We have much to learn from those peoples who live sustainable, relatively self-contained lives in harmony with their environment.

A community government will need to track the outside world, to watch trends and suggest changes and ensure that the outside world knows about the goods and services it can supply. It needs to ensure that the organisations that are needed to sustain its population and its relationship with the outside world are present and thrive. Therefore surely the first necessity of sustainability is to ensure that control is not passed to the outside by allowing outside ownership of community assets? This seems to be the opposite strategy to that which government agencies that I have interacted with have pursued for the whole of my lifetime. The mantra has always been that 'we must attract outside investment', that is give away our control. No wonder 'take back control' proved to be such an attractive slogan in the UK. A community must achieve a balance: allowing in new ideas and learning, but not allowing a critical loss of control. The second necessity of sustainability is to have the institutional ability to understand the community's relationships with its environment, both social and ecological, and the ways in which it must maintain the knowledge and skills on which its trading relationships rely.

We need to be clear that in considering government for sustainability there are dimensions to government that are now substantially missing from our democratic institutional structures. At present these institutions are organised to cover social aspects of governing, but not the skill or ecological aspects of governing, which are required to develop and maintain inter-community trading relationships, and right relationship to the ecosystems of the earth.

8.4 Constitutional Guidelines

In Chapter 4 I explored the consequences of moving to modelling an organisation in a systemic way starting from considerations of holding steady in a changing environment and having to deal with both internal and external disturbances. This approach resulted in a layered fractal structure. That layered fractal structure is of course the reason for fixing on the word 'community' because it is a word which is essentially fractal in nature, in that it is used to describe a system in focus as is required for a systemic analysis. Figure 8.4 reproduces Figure 4.11 now illustrating the layered structure of a community government and its relationships to its environment and to the sub-communities and sub-sub-communities.

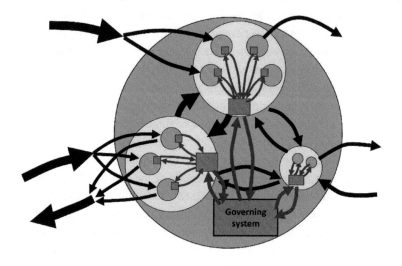

Figure 8.4: The Layered Community Structure

Many governments have evolved to have a layered structure, but it has to be said that the relationship between the layers does not seem well understood. In the UK in 1977 a Scottish Member of the United Kingdom Parliament, Tam Dalyell, posed a question that is still a subject of controversy a generation on; it was dubbed the 'West Lothian Question'. The question asks: should Scottish Members of Parliament vote on issues concerning only the governing of England? This question has been generally regarded as having no answer, and perhaps that is the case for an object thinker. But it is exactly the question of how adjacent levels of management should relate and this was resolved systemically in Chapter 5.

As is described in Chapter 5, all team games solve this problem. To recap, in any team game there is a set of rules which apply to all the players taking part in a game, the metasystem. Within this agreed governance framework each team and its players have the freedom to play their game as they wish and take their own team decisions among themselves. The rules of any game give rights to players, 'freedom to' act and communicate in certain ways, but also provide constraints curtailing freedom, giving players 'freedom from' being on the receiving end of other particular acts and communications. A team can on its own account set higher standards of behaviour, but not lower, and they can certainly decide their own strategies of play. This relationship between the whole group rules (the framework which establishes the game) and the subgroup rules (the strategies that a team uses in taking part), and the way in which the two levels relate, is an example of governance in the form needed to provide the answer to the West Lothian question.

The whole community government is concerned only with this whole group framework and the issues associated with its establishment and maintenance. In normal circumstances there can be no concern with internal matters of any sub-community, just as is the case for the sovereignty of nation states within international law. The key to understanding the relationship between two levels of government is that there must be restrictions on both the upper level and the lower level on what can be decided, what legislation can be put in place. There is inevitably a greater diversity of views across the larger group than across any sub-community. The area of agreement on a framework will of necessity be on a minimal set of agreed rights and constraints. The decisions of the whole community government must be restricted to those affecting only the whole community in constructing and maintaining this framework.

In the example of England and the whole of the United Kingdom, issues concerning England only should be considered and decided by a body or bodies governing England: an English parliament or appropriate regional parliaments. This (or these) must be sovereign alongside the Scottish, Welsh, and Northern Irish parliaments within the common United Kingdom framework set by a United Kingdom constitution. With this principle as the starting point in designing a governing structure, the problems we face in that design are enormously simplified because the principle can be applied to distinguish the functions of any two adjacent levels in a multi-level governing system.

However, in adopting this principle we consign to the medieval history books the principle of sovereignty as it is currently defined and used. At present the government of a larger community can interfere arbitrarily in any way at any time it deems necessary in the smaller contained community. In the past sovereign kings and queens have not been renowned for their open, informed, fair and just decision making. The notion that there is a hierarchy in which the higher echelons can interfere at will in the lower echelons must be abandoned. Powerful and unconstrained actions in the upper echelons of a hierarchy certainly do not give rise to open, informed, fair and just governance, but more likely give rise to moral and financial corruption. In my experience it seems that those in the higher levels of a hierarchy often seek to perpetuate their own position and power rather than govern as representatives of the population.

What do we gain from defining the relationship between levels in this way? At one extreme of decision making it is clearly a fantasy that a king, a queen, a pope, a prime minister, or a chief executive officer can be infallible, and all-seeing. All human beings have a limited capacity to understand the enormous complexity of the world in which we live, whatever their role in it. They do not have the necessary variety of potential command. The collective knowledge of the governing body of the larger community containing the smaller is limited. Even a few members representing a city in a national government living for a large part of their time at some distance from that city can never act collectively as intelligently as a local city government body of many more people in everyday touch. So in structuring any governing system, it is as essential to prevent interference from those who are not affected and do not understand, as it is to give access to those who *are* affected and *do* understand.

Narrowing the domain of decision making in this way will focus the minds of those involved in the whole community government on a range of issues and roles which can be better encompassed by a human brain, so increasing both the effectiveness and efficiency of governing structures. This at least maximises the possibility that in any government the people affected and the people who understand are both involved as far as possible in the decision-making processes which affect them. This approach to structuring will tend to decentralise decision making, since unless it can be argued that a decision affects the whole community group, it cannot be considered by the government of the whole community group.

8.5 A Systemic Approach to Democracy

The first example of democracy, and the origin of the word, is from Greece. In the city state of Athens in ancient Greece all citizens took part in governmental decision making – what we now call *direct democracy*. In ancient Greece women and slaves were not citizens, and therefore not directly involved, but in the way we think of it now, in a *direct democracy*, every adult is a citizen. However, who is or is not a citizen even in our modern world is not a simple question for any community; there has always been movement of people from community to community spreading knowledge and understanding. When does an incomer become a citizen able to take part in the governing system? What qualifications should there be?

It makes sense to bring to bear on decisions all the available knowledge and understanding of the issues involved that exists in a community, and also all the available understanding of ramifications of any decision, how it will impact on the members of the community. This is essential if we wish to have the best possible model in a community control system taking note of the Conant-Ashby Theorem. But central to this purpose is that there must be built into the governing system the necessity for there to be discussion to bring together the different perspectives which will exist on any situation. From that discussion agreed solutions to the complex problems of governing will emerge. As outlined in the previous two chapters, each citizen has a unique personal set of experiences which give rise to their view of the world. On any issue there can be a variety of views, both informed and uninformed, and in order to reach a decision there must be that time for open discussion to enable ideas to be challenged, misinformation to be countered, and a decision to emerge. In this process great care must be taken to prevent any group even a majority of citizens curtailing the discussion to impose its will without that informed discussion. John Stuart Mill writes about the necessities and problems with both the informal system of societal customs and the formal legal system in constraints on personal action in his writing *On Liberty* (1987) including:

> ... 'the tyranny of the majority' is now generally included among the evils against which society requires to be on its guard.
>
> (Mill 1989)

A House of Representatives

The problem we have in making a comparison between the government of ancient Athens and modern communities is that, from my definition, communities can be very large – up to and including the community of humans on planet earth. In Athens every citizen could be involved in governmental decision making, but in a nation state with a typical population in the millions or tens of millions this is not feasible. In the modern situation it seems to be generally accepted that *'representative democracy'* is the ideal to aim for. Instead of every adult being involved, we choose representatives, creating a body to take decisions on our behalf. The meaning of to 'represent' in this context according to the dictionary (e.g. Chambers 1998) is to 'stand for'. The purpose of this 'standing for' is that the representatives take governing decisions on behalf of the people of the community as a whole, to the benefit of the whole community. If representatives are going to *stand for* the people of the community as a whole, then they must carry with them to their task, as far as possible, the rich variety of background, culture, and attitudes of all the people of the community which they represent. Culture varies from place to place, from organisation to organisation, and from profession to profession; a governing system must capture as far as possible this variety. If representatives are taking decisions standing for a community of people then there must be as far as possible an alignment with, and understanding of, this rich variety in that controlling body. Only then can they *stand for* the population in deciding what is appropriate, and what is not, making those decisions to the benefit of the community as a whole.

But then the question arises as to the ability and desirability of a small number of people to represent a large number in this way. What is then at issue is the quality of those decisions since all citizens are not involved. The fractal constitutional structure explored in the last section does alleviate this problem, restricting the range of decision making at any level of community, but it does not entirely solve the problem. The implications of the Conant-Ashby Theorem enjoin us to seek to establish in the controlling governmental system the best possible model of the community and its needs and problems. How could this be done?

The VSM – the systemic control system proposed by Beer, explored in the previous chapters, and illustrated in Figure 8.5 – is divided into five sub-systems. System 5 is to embody *the being* of the whole system, and is the guardian of the identity and purpose of the system as was described in section 4.7. An animal is itself, its *being* developed by coevolution with the environment within which it lives. In the modern world a community has the opportunity to develop an understanding of its being and decide how it wishes to relate to the environment. In a direct democracy this is the responsibility of the whole collective of the citizens, in a representative democracy the responsibility of a *House of Representatives*, standing for the whole collective of the citizens. As guardian of this identity and purpose, the function of a House of Representatives is to decide what is appropriate and what is not in proposals it receives and to make decisions for the people of that whole community, but most importantly, the fractal point, it can only make decisions which apply to the whole community. I explored examples of what might be the identity and purpose of a community in section 8.3 at the beginning of this chapter.

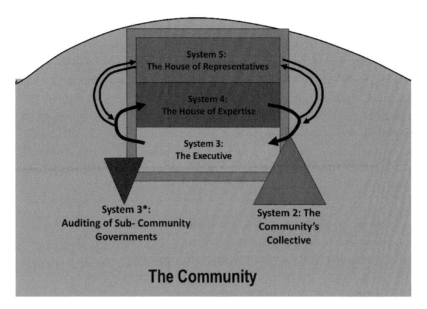

Figure 8.5: Community governing systems

It is essential that there is understanding in the House of Representatives of the rich variety of knowledge, understanding, and perspective in the community. Any democratic government therefore **by design** consists of groups of people representing their community who come from different backgrounds having differing perspectives on the complex situations they face, that they must come to an understanding of, and come to decisions about. The decision-making process of democracy requires inclusive discussion, which requires developing agreed models of complex situations as discussed in Chapter 6. This is not at all the way our current governing systems are structured. It is more usual for governments to strip out the variety of perspectives, UK political parties are typical in that the choosing of parliamentary candidates to stand for election as 'representatives' has been decided in large part centrally to fit with a particular political view. This is done often even without the candidate having any experience of the constituency concerned; a sure way to achieve incompetent governing. To achieve competent governing it is necessary to maintain an ongoing feedback loop between a House of Representatives and its community to ensure evolution with the community, and to counter misinformation and misunderstanding. To achieve this I would suggest also that frequent changes of representative are necessary to counter the possibilities of groupthink within the House of Representatives; representation cannot be a permanent career.

The House of Expertise

The present system of representation has hardly changed from the time when nation states were largely rural farming societies, when the differences in working practices across nation states were relatively small. Then a division of the nation state into geographical areas could by and large encompass the knowledge and understanding across that nation state. That has ceased to be the case for some considerable time. A geographical constituency structure is no longer adequate to produce a representative understanding in a governing system. In the *New Scientist* magazine of the 24th April 2010 Michael Brooks reported:

> In the last parliament, only 27 members out of a total of 650
> held a science degree; 584 members described themselves as
> having no political interest in science and technology,
> according to figures from political research organisation Dods
> People. Some ... are positively anti-science.

Without interest there can be no understanding, and yet understanding of science and technology is clearly a must for viable government in the world in which we find ourselves.

Understanding within the governing system could be vastly improved by having representatives of expertise areas. It seems essential to suggest that alongside any *House of Representatives* based on geographical communities, there also should be a *House of Expertise* and this requirement should be a permanent part of the structure. Such a House of Expertise should be elected from communities based on skill and knowledge in working practices. This would be a second fractal structure in which all citizens play a part. In today's world it is not difficult to generate a set of constituencies covering the range of skills present in any geographical community and each and every citizen of that community to be registered and to vote in one of those skill-based constituencies, just as we do now for the geographically-based constituencies. It seems logical to encompass as far as possible all differences in our representative structure. Trade guilds and trade unions are examples of organisations based on skill areas that have existed for many years.

The House of Expertise is the community managing System 4 (Figures 4.13 and 8.5) as proposed by Beer. Its purpose is to monitor the environment of the community, learn, and propose changes that it feels are necessary for the survival of the community. It must monitor all relevant aspects of the external environment – ecological and social – maintaining necessary external dialogue and trading relations, and proposing changes necessary for survival of the community. Of necessity, therefore, it should embody all the skill areas of the community and be a centre of learning and innovation. It is *the higher brain* of the community where learning and adaptation originate. Therefore, an integral part of the function of the House of Expertise is the maintenance of the education system of the community. My experience of taking part in the governing of secondary schools led me to the conclusion that an educational opportunity is missed in that students are not engaged in the analysis of the needs, current and future, for the maintenance of the community in which the school is situated. Similarly students of the primary and the tertiary education system could be performing the same function for their communities at the appropriate fractal levels.

A 'citizen's wage', sometimes labelled a 'universal basic income', has an educational role to play in enabling time and space for creativity and entrepreneurship, those not immediately engaged can reskill to provide for coming changes, or be available when there is a sudden demand increase signalled from the Community. Such a scheme would maintain redundancy in a community, just as all bee colonies maintain redundancy in order to cope with changes in their environment. Bee colonies are organised in this way so that they can and do operate sustainably.

The Executive

The third sub-system of the community governing system is the *Executive*. In Beer's VSM this includes System 3, System 2 and System 3˙ (Figure 8.5). The purpose of the Executive is to engage and manage the sub-communities, playing the role of *the lower brain and nervous system* of the community. Its overall purpose is to maintain the operations of the system as discussed in Chapters 4 and 5. The most important of these is System 2, because of the complex nature of maintaining internal coherence and stability across sub-communities. The tasks of this part of the Executive cannot be encompassed by a centralised body and must of necessity be operationalised by both the government of the whole community and the governments of the sub-communities. This I have designated as the responsibility of a *Community Collective* (Figure 8.5), which is a body consisting both of representatives of the sub-communities and representatives of the whole community and which will be discussed in the next section.

System 3 itself is a system meta to the governing systems of the sub-communities with the purpose of maintaining a whole community view of the activities of the sub-communities. As discussed in Chapter 5 and 6 it allocates to those subsystems the resources that are required to meet particular challenges from the external environment. This would include building and putting into action new organisations, in the same way as putting resources into building and putting into action new learning in any animal. My learning of music and piano playing, referred to in Chapter 6, is such an example: the creation of new understanding and new skill. The last part of the Executive responsibility is of *audit,* System 3ʾ, to ensure that the picture presented to the Executive of each sub-community by that sub-community's House of Representatives is an accurate one. These mechanisms with those of the Community Collective constitute the feedback loop between the whole community government and the sub-communities.

The structure of Beer's Viable System model envisages a continuing discussion between the Executive and the House of Expertise informing each other of the state of viability of the community. Are the efforts of the citizens of the community, the inside and now (the current understanding in the Executive), aligned with the future evolution of its environment (the current understanding in the House of Expertise)? If change is necessary then agreed change, emanating from the House of Expertise's learning and creativity, and the ongoing discussions between these two bodies, is proposed to the House of Representatives for their decision – are they in line with the culture and ethos of the community? These ongoing conversational feedback loops are illustrated in Figure 8.5 by the curved black arrows. The House of Representatives is the decision-maker, it sits, as required by our understanding of control systems, meta-systemic to the ongoing Executive – House of Expertise discussion. The Representatives are therefore required to be meta – that is outside, above – both the House of Expertise and the Executive. Understanding of these roles is vital for good government. At the larger scale I would propose that one citizen could not be a member of more than one of these three subsystems, House of Representatives, House of Expertise, and Executive, but this may not be possible at the smaller scales, e.g a small village.

Given the complexity of governing tasks (it should be no surprise that it is complex) it does not seem unreasonable to require that any potential representative or executive appointment undergo professional development, as in any other profession that requires decision making in complex situations. This is an accepted normal process that applies in such professions as medicine, the law, architecture and others. If a person aspires to be a manager of any organisation then they require both the specialist knowledge of that organisation and knowledge of managing, both sociological and cybernetic. The route to representative must be open to all, which will require some thought into how the necessary understanding, openness and true representation can be achieved. Perhaps there should be a rule that a citizen could not be a community representative without having been a sub-community representative, and terms of office limited to prevent representatives becoming separated from their community; all these being fractal requirements in line with the fractal systemic structure of governing, but allowing many to participate. Choice by sortition rather than election seems attractive, but if by election, certainly proportional representation is a must to ensure that multiple perspectives and experience are present.

The Community Collective

In sections 5.4 and 5.5 I explored achieving coherence between a group of sub-systems, applying the Sub-optimisation Theorem, the last of the three cybernetic rules which apply to all managing situations. It shows that in order to optimise any variable for the whole system, sub-systems must be prevented from optimising that variable for themselves in isolation from other sub-systems. This is achieved by imposing a frame within which they must work which enables the co-operation necessary to bring coherence to the whole. For example, it is well understood in the world of soccer that having a prima donna in your team does not lead to a high performing team. But, it does not seem to be generally understood in the world of government that in states with high inequality both economic and social development are damaged (Piketty 2020). It is to be expected that if organisations or sub-communities are allowed to maximise their own well-being then overall well-being is damaged.

For any community the task of preventing sub-optimisation splits into two parts. The first is maintaining the structure of the community: that is maintaining the economic balance between the sub-communities, enabling necessary change in boundaries between sub-communities, maintaining cohesion through trading relations, building new organisations for a changing environment. The second part is the development and maintenance of the underlying framework of rights and responsibilities that applies to all people and organisations within the community. These two ongoing tasks and their interlocking nature are illustrated in Figure 8.6.

In Chapter 5 I used government as the principal example in showing that a whole community must develop and manage a minimal frame within which sub-communities at the next level must adhere if it is to avoid the problems of sub-optimisation. Each sub-community can add additional constraints to the minimal frame but not take anything from the constraints imposed.

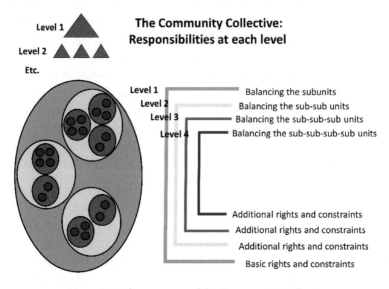

Figure 8.6: The purposes of the Community Collective

Government creates laws which describe rights and constraints. There are some things that I, any person, or any organisation can do and some things that they can't without sanction, some things that can be done to me, any person, or any organisation and some that can't. It is this that forms the frame of acceptable things that can and can't be done. This frame deals only with things of relative importance, for example it says that I can't take things for my own use that others habitually use, and I shouldn't kill anyone. They protect me from having my tools stolen, or from being killed. The regulations act in both directions, limiting how I affect others and how others affect me. They don't say that I should only use a circularish spoon for eating soup rather than an oval shaped one, although such cultural regulations do exist. So in any society there are some regulations that are informal, and some that are formally enshrined in law, but all change and evolve with time. As Piketty (2020) argues, the framing within which we live our lives is largely constructed by the stories we tell ourselves. There are always many more options for developing the formal legal structures than those which are adopted, and they are adopted because they seem to be the obvious solutions dictated by the current framing narratives.

The simplest part of this aspect of governing is that it is a system to regulate relationships between citizens, or perhaps to constrain the behaviour in relationships to those considered acceptable. The formal part of this is the legal system. As discussed in section 5.5, the whole community frame is a minimal set of regulations which apply to all sub-communities. The important point is that in the framework there can be no discrimination for any reason between individuals within a whole community, but each sub-community may add to those regulations according to their custom and practice as long as the whole community regulations are not breached. This gives the structure of the fractal layered framework. This is illustrated in Figure 8.7 reproduced from Chapter 5 Figure 5.6. The colour represents the framework at a particular level, thus the common frame for the whole community is represented by the white colour. For each step to a sub-community a small amount of colour is added to illustrate the additions to the frame. It can be seen from the illustration that it is possible for communities with very different cultures and ways of living to exist in such a fractal structure.

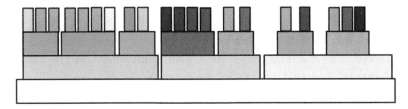

Figure 8.7: An illustration of the layered structure of the internal governing frame

The situation for a government is even more complex. Whilst a government is concerned with relationships, it is not just those between individuals. In any community there are, in addition to relationships between its members, relationships between organisations in the community, and relationships between members and organisations. Just as in many ecosystems, organisations are present in some communities that also operate at a larger community level. In the ecology of the planet, a large tree, or a top predator operates across many ecosystems and modifies or even creates ecosystems within its influence range. There are many business organisations which span geographic communities and even create geographic communities in similar ways. Employees and supplies are drawn from a range of communities, and products and services are also distributed and used across a range of communities. The three ranges are not necessarily the same. In our current world, nations suffer from business organisations playing nations against each other; this points to a need for a world frame. The agreed frame across sub-communities at all levels is of utmost importance to eradicate this behaviour, and this leads to the conclusion that the frame governing the standards which apply to an organisation must be set at the level that encompasses all its operations.

These frames are control systems as I discussed in Chapter 5. The first aspect of the community control system is social regulation of relationships. This could be someone detected behaving badly and being told, and losing friendship, or an organisation behaving badly and losing custom. There are many examples of ways in which people themselves act as sensors, comparators and actuators in changing their own behaviour in an endeavour to put pressure on miscreants to change theirs. The formal system, that is the legal system, is illustrated in Figure 8.8. The formal system picks up the problem when a matter is serious enough to be reported to the police or other regulatory body, who then have the duty to investigate and bring a formal charge of transgressing the legal framework. The court system has the duty to compare the model of the situation produced by the investigation with the accepted framework standards. If it is found that the framework has been transgressed then sanctions will result. These sanctions are of little value if the behaviour of the transgressor does not change. Therefore the sanctions must include the purpose of changing the models that guide the behaviour of the miscreant. This is the actuator which brings about the change and corrects – or should correct – the bad behaviour.

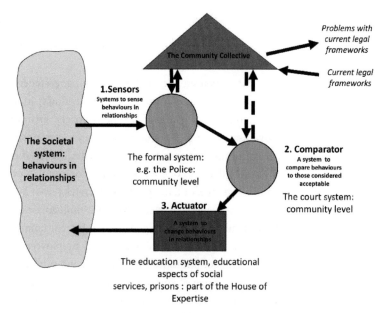

Figure 8.8 The formal community regulation system

As a society evolves, so must the formal legal system. When a problem is perceived with the operation of the current sub-communities framework, the Community Collective has a duty to bring problems to the ongoing discussion between the Community Executive and the community House of Expertise. It is important here to remember the levels' relationships, the community sets the frame for its sub-communities and may change that frame. A community cannot change the frame within which it sits but can of course bring problems to the attention of the meta-community collective. Following this logic, in cybernetic terms and in terms of the proposed democratic system, the necessary control system to effect changes in a sub- communities framework can be modelled as in figure 8.9.

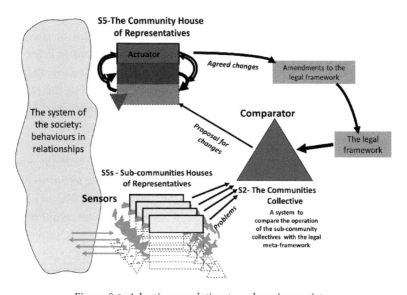

Figure 8.9: Adapting regulation to a changing society

Beer's conception of the Viable System Model (VSM) proposes that there is ongoing feedback communication between System 2s – i.e. the Community Collectives – between levels. Therefore the collectives themselves form a fractal structure. This fractal structure will include the operational subsystems of the Community Collectives, that is the court system and judiciary on the one hand, and the police and other behavioural formal sensors on the other. Both these should also be fractal structures but needing further analysis to see how far from that they are now, and of course they have their own governing frameworks.

8.6 Conclusion

The situation that Forrester describes in the quotation at the head of this chapter has changed little in the years since 1970, although now Piketty has gathered a considerable weight of evidence on the practice that has been tried and suggests in his book possibilities to investigate (Piketty 2020). I suggest that applying systemic logic should also be a guide on the path to better governing. A governing system is a system to regulate relationships. Relationships are the interconnections between things, very much the subject matter of Systems Thinking. Governing is a process of holding steady in a changing world, and the subject matter of cybernetics, exactly as both Plato and Ampère envisioned. What is surprising is that this approach seems entirely absent from the literature on governing, even from most current writing. When I analyse the workings of the United Kingdom government, the one I know the best, it accords to none of the three cybernetic laws, nor with the best practice of ensuring the well-being of all citizens or achieving right relationship to the natural world. Human society and its relationships to the natural ecosystems of the planet is a hugely complex system, therefore it seems to me that any attempt to analyse and discuss governance in any conceptual system other than a process systemic conceptual system is doomed to failure.

Most of us would, I imagine, wish to live in peace, without harming others or being caught up in the stories that competing elites create to further their position, and being able to get on with our chosen lives. We need the communities in which we live to survive and be reasonably prosperous so that we can maintain our own livings. Those we elect to positions of governing, our representatives, have firstly the task of living in and listening to the society they represent. Societies change over time, and the stuff of government – constitution, laws, regulations – must change with it. Representatives have the task of using their knowledge and understanding to translate the changes into amendments to the stuff of government. This stuff of government should ensure that we can live in peace, without being harmed by others, and help maintain all the members of its communities. At root then, the task of our representatives is to ensure that the values of the society are reflected in its laws and regulations, and that all have the possibility of a decent living. This is the purpose of managing a society.

9. Modelling Quantum Systems

9.1.   Introduction

My own starting point in research physics was in January 1966 when I was appointed to a research position in the Physics Department of Royal Holloway University of London and began research for a PhD. I was attached as a theoretician to a group of experimental physicists who were researching the physics of the laser, which was at that time a new device. This PhD, completed in the summer of 1968, was my first foray into the theory of the way in which matter and light interact (Asby 1969, 1970).

In September 1968 I moved to work in a group in the Department of Physics and Astronomy in the University of Rochester, NY and became fascinated with the general problem of the interaction between matter and an electromagnetic field. The standard method followed in the derivation of the equations modelling this interaction uses a standard reductionist approach. This standard approach to electrodynamics assumes that it is possible to change separately each of the two parts of the system, the electromagnetic field and the electrons, each without affecting the other. This standard approach reaches an impasse and only with a dubious workaround of this impasse can it be completed. This problem with the approach indicated to me that the reductionist assumption was not valid. I found support:-

> A Problem with electrodynamics
>
> I feel that it is a delusion to think of the electrons and the fields as two physically different, independent entities. Since neither can exist without the other there is only one reality to be described, which happens to have two different aspects; and the theory ought to recognise this from the start instead of doing things twice.
>
> (Albert Einstein)

I then tried a new approach without the reductionist assumption, but this did not overcome the problem. However, what fascinated me was how the pattern of the calculation turned out (Asby 1973). Unfortunately at this juncture my time in research physics came to an end, though my interest remained. As far as I am aware this avenue of research has not been pursued further.

After eight years of employment in education, as a result of my interest in government I first met Stafford Beer. At that first meeting conversing on our joint interests, he made the conjecture that Quantum Mechanics was an application of Systems Thinking. Whilst intuitively agreeing, on reflection I determined that I needed to improve my understanding of Systems Thinking in order to definitively answer this rather fascinating conjecture. My interest in Quantum Mechanics continued and ten years ago or so I came across the work of Bernard d'Espagnat (2006). In this work there is a comprehensive analysis of the problems of Quantum Mechanics, and d'Espagnat reviews the various interpretations of the mathematics. He concludes that Quantum Mechanics is a process theory: a process being the fundamental entity on which the models of Quantum Mechanics are based. This conclusion agrees with Beer's conjecture when it is remembered that a system is defined as a purposeful input→process→output structure. Whilst d'Espagnat comes to no conclusion as to any particular interpretation of the mathematics, this finding added impetus to my quest.

Following my meeting with Beer I focused on the development of my understanding of Systems Thinking, but also researched a means of representing Beer's fractal systemic approach to Systems Thinking in mathematical terms. It is the synthesis of these three strands of research – approaches to understanding interacting systems in physics, the conjecture that Quantum Mechanics is a systems theory, and the mathematical modelling of systems – that have resulted in this chapter. The purpose of the chapter is to show that the approach developed in the rest of the book does verify Beer's conjecture and also leads to a possible new understanding in the interpretation of the mathematics of Quantum Mechanics.

9.2 Flaws in the Newtonian Approach to Modelling in Physics

The foundation of WEIRD thinking lies in the philosophical approaches of Descartes and Newton. The WEIRD approach led to great successes in physics – it is an approach that assumes that an understanding of a phenomenon can be obtained by seeking to model constituent parts of the phenomenon, then putting those understandings of the parts together will give an understanding of the whole. This can be traced back at least to Democritus who is credited with being the first person to formulate the idea that matter was constructed from atoms as a fundamental particle. The WEIRD approach further assumes that the models obtained fully represent, or even are, the phenomenon. It is both reductionist and realist. However, around the turn of the twentieth century, serious problems arose with this approach. Neither the understanding produced by the mathematical models describing the two Theories of Relativity, nor that of the Theories of Quantum Mechanics which were proposed to explain experimental anomalies, fit into this original framework.

In Newtonian theory, measurements of time and distance are taken to be independent of the observer, as is expected in an 'objective' approach. However, relativity theories proposed that the measurement of time and distance depended upon an observer's perspective on the situation under scrutiny with respect to relative velocity and gravitation. For the first time it was necessary to take into account observer perspective in understanding the results of observations of physical phenomena. Whilst this was a revolution in physics thinking at the time, subsequent experiments to test the theories verified them and, as a result, the new ideas introduced by the two Theories of Relativity gradually became accepted.

The situation with Quantum Mechanics is different. The results of experiments to test the mathematical modelling of the theory have produced results verifying the mathematical models to unprecedented accuracy. But it has not proved possible to come to agreement on how these results should be interpreted. A number of different interpretations of the mathematics of Quantum Mechanics have been proposed, and there are still unsolved philosophical problems. Much argument and discussion has taken place over how to understand the quantum mathematical modelling over the many years since it was first proposed. The WEIRD view is challenged in two main ways, the first being, 'what do we mean by reality?' and the second being, 'how do we model interacting systems that naturally lead to collective behaviour which cannot therefore be localised?'

Chapter 6 developed a systemic model of learning which provides an understanding of Scientific Method as a collective learning process. This formalising of the learning process was discussed in Chapter 7 and resulted in Figure 7.5 reproduced here as Figure 9.1.

Scientific Method introduces the key ingredient that the models proposed by an individual must be testable by experiment, and are expected to be published in order that others may also test the model. If a model survives such testing then over time it will be adopted by the relevant scientific community as a viable model and may eventually be regarded as 'reality'. The observation of the milestone that I described in the last chapter was a personal experiment resulting in the perception of a pattern of signals which must be interpreted. I argued in Chapter 2 that models underlie all perception. Underlying the perception of the milestone pattern are models built into the human perceptual apparatus, and learning gained from early childhood experiences as described by Piaget (1954). I approached the milestone, and as I changed my perspective the perceptual image behaved as I expected. Coming close, I could then touch the surface, walk round it, testing and verifying my model which predicted a stone cylinder. Testing models is a fundamental step in the learning process which has been formalised in Scientific Method as a necessary step in creating a model. An experiment, such as my observation of the milestone, is a system, the purpose of which is to test a model, and the output of which is an observation, or set of observations.

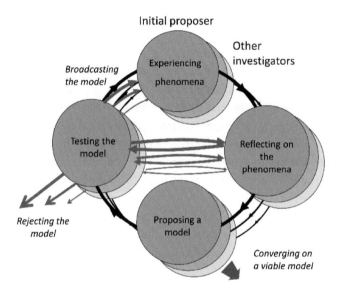

Figure 9.1 Scientific Method

The model predicts the expected output, which, if obtained from the experiment, is one step in verifying the viability of the model. However, if the predicted result is not obtained, then the model is refuted and should be abandoned (Popper 1972). In any experiment, including one designed to explore the world of quantum systems, the apparatus itself is a system, a particular structure designed to produce an input to the system of interest under study and an apparatus to record the output from that system: Figure 9.2.

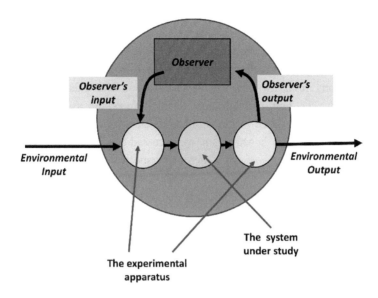

Figure 9.2: The experimental process

The key point that I wish to stress is that whatever is observed is observed from the singular particular perspective that the experimental apparatus dictates, and the model that guided the construction of the apparatus predicts the observational outcome. The notion of perspective has immediate impact in interpreting experimental outcomes. I take as an example the interpretation of the results of the 'double slit' experiment (Feynman, Leighton, and Sands 1963) wherein electrons, originally modelled as particles of matter, produce the interference patterns of waves. It was shown that even individual electrons appeared to pass through both slits of the experimental apparatus. But if the experiment was modified, by placing a detector to ascertain which slit the electron passed through, a different result was obtained. The wave interference patterns disappear and now the electrons behave as particles would be expected to behave. From a systemic viewpoint it is not a surprise that a different experimental system with a different structure produces a different result. From the general principles developed in Chapter 3 it follows that although the input is the same in each of the two experimental situations, the structure of the process is different, and, as a result, a different output should be expected. The observations made, the results of the two experiments, are two different perspectives on the situation under examination.

Most of my observations of Roman milestones gave me the result of a tall rectangle standing up from the ground. A small number gave me the result of a circle, a wheel resting on the ground. Others were more complex; the result of each experiment gave only a partial view of the underlying situation under scrutiny. If there are different configurations of the experimental set-up giving rise to different experimental processes, then I argue that these different configurations result in different perspectives on the phenomena being investigated. The first configuration clearly refutes the assumption that electrons can be modelled in terms of the classical concept of a particle, an object which is itself a construction, as discussed in Chapter 7. It is much better to think in terms of two different perspectives on a more complex phenomenon.

Bernard d'Espagnat (2006) describes the notion of a particle in Quantum Electrodynamics as follows:

> ...the existence of a particle is a state of a certain "Something",
> that the existence of two particles is another state of this same
> "Something", and so on. Of course, the absence of a particle is
> also a state of this "Something". Then the creation of a particle
> is nothing else than a transition from one state of this
> "Something" to another.

The question is: what is this something?

A systemic approach founded from a first-person, subjective Pragmatist perspective also removes other difficulties of Quantum Mechanics. An example of that is the famous conundrum of Schrödinger's cat, in which a cat is enclosed in a box with a cyanide capsule whose contents are released by the decay of a radioactive atom. The decay of the atom occurs probabilistically and cannot be predicted. An observer outside the box can only ascertain the state of the cat by opening the box. Whilst the box remains closed the cat is alive/dead with a known probability. At this time of writing I am unsure of the whereabouts of my own cat. He could be in a number of different places with certain probabilities ascertained from the data I have gathered on his habits. When I search I find him in one of those habitual places. In this case, and in the case of Schrödinger's cat, the ultimate observation changes the probability to certainty. Someone sitting alongside my cat has a different perspective.

A human observer of Schrödinger's cat – inside the box, suitably masked and observing the cat – would observe the moment of death. From the systemic point of view there are as many worlds as there are observers, as many worlds as there are living entities. Each living entity inhabits a different world of internal models built from its genetic inheritance, and if it is capable of learning, its lifetime experience. This is the systemic conclusion, that the problem of Schrödinger's cat is a problem only for someone with a realist worldview.

As a result of the exploration and discussion in Section 7.2, I summarised the purpose of internal systems (process) models and concluded that:

> An internal model is developed in order that an animal can predict the state and configuration of the environment in the future, given the state and configuration at a given moment, from the viewpoint of that animal at that time in carrying out its projects. The models enable the animal to achieve its necessary functions in its econiche. In life there are many surprises, so a model will not guarantee survival but as long as it enhances the probability of survival a particular model will persist in the animal's repertoire. The purpose of a such a process model is to predict how the environment flows, not what things are.

This claim, that the purpose of the model is to predict the configuration of the environment at a future time, is fundamental to the way we think and model our environment. It fits exactly the experimental approach described above and in particular the approach that must be taken with Quantum Mechanical models.

> ...what most basically differentiates quantum mechanics from classical physics is not (as often believed) the fact that its axioms involve intrinsic probabilities. It is in fact that it is not descriptive but essentially predictive, and, more precisely predictive of outcomes of observations.
>
> (d'Espagnat 2006)

This is the fundamental point of connection between the Systems Thinking approach developed in the previous chapters and Quantum Mechanics. It confirms Beer's conjecture made more than 35 years ago and my intuitive acceptance of that conjecture. Thinking of modelling in this way is fundamental to the systemic approach, and to the way in which observations from experiments should be treated, and further, of course, to the approach to Quantum Mechanical models and modelling.

9.3.   Qbism: The Quantum Bayesian Approach

Having come to these conclusions, I was surprised to find that this systemic view is exactly the Quantum Bayesian, or Qbist, view (von Bayer 2016, Fuchs 2017, Ball 2018). Qbism makes the assumption that:

> ...in Qbism ... all quantum mechanics refers to are beliefs
> about outcomes – beliefs that are individual to each observer.
>
> (Ball 2018)

This Qbist view is obtained by being rather careful how the probabilities within the mathematical models of Quantum Mechanics are interpreted. In the QBist approach those probabilities are understood as Bayesian probabilities. That is, they are strictly associated with the observer experimenter, attached to their perspective exactly as the systemic approach would dictate. Each observer has a unique personal perspective depending on their past experience and the information that they have of the experimental situation. This is in contrast to the WEIRD approach to the mathematical models of Quantum Mechanics where probabilities are understood to be 'real', a statement of the likelihood of a particular outcome as a realist would require. Hence the Schrodinger cat thought experiment produces a paradox, how can a cat 'really' be both alive and dead and then suddenly be either one or the other when the box is opened.

The final point in common that the Qbist approach has with the systemic process approach is that it is founded in Pragmatist philosophy (Fuchs 2017) just as I discussed in Chapter 7. Therefore, my conclusion is that the foregoing chapters provide a systemic foundation for the Qbist interpretation of Quantum Mechanics. The Qbist approach now appears naturally as a particular application of the general systemic process approach to modelling the natural world not limited to quantum phenomena. Secondly, the systemic process approach enables a new suggestion for a different way forward as will become apparent in the next sections.

9.4. The Importance of Symmetries

I explored in section 7.9 the way in which we construct 'object' models by viewing objects from different perspectives and synthesising the results of observations. In the process of observation, symmetry plays a fundamental role, which can be illustrated by further considering my observations of the Roman milestone. I said that most of the observations that I made of a milestone produced a tall rectangle. This is perhaps stating the obvious, but it arises from the fact that the milestone is in the shape of an upright cylinder which has rotational symmetry. As a result, any observation from the side produces the same result. A moment's reflection on this gives a perhaps surprising but extremely powerful insight. If I observe any situation from a given particular perspective, and obtain measurement results, and if that situation under observation has a symmetry, I will obtain the same results from approaching the situation from any perspective that is symmetric with this first perspective. Thus, in this systemic approach symmetry is a fundamental consideration in carrying out observations of any situation under scrutiny. Symmetry is not in this consideration a symmetry intrinsic to a situation under examination, but a symmetry intrinsic to the whole, observer plus situation. The observer's relation to the situation under examination is an integral part of the whole. This insight is therefore fundamental to Scientific Method. Two different experiments exploring the same situation and giving the same results indicate a potential symmetry; two experiments giving different results indicate two different perspectives on the situation as with the two-slit experiment discussed above.

In 1935 Einstein, Podolski and Rosen published a paper questioning the interpretation of Quantum Mechanics. The EPR problem (as it became known) became the focus of much subsequent attention and gave rise to the concept of entanglement to describe the situation where two or more particles, in object conception, form an indissoluble system. Bernard d'Espagnat (1999) describes a classical analogy of the situation as:

> A car equipped with two axles is sent into free space. Each axle has previously been given a spinning motion, and the angular momenta of the axles are equal and opposite. At some time an explosion takes place in its middle and tears the car into two pieces, front and rear. ...Observers can subsequently make measurements of various components of [axle spin] defined in a given reference system. They may also compare their results in order to investigate the correlations.

The spinning axles move away from the site of the explosion in opposite directions. Observer A and observer B observe the explosion from opposite sides, each looking towards the site of the explosion and both see an axle moving towards them, and an axle moving away from them with equal and opposite spins, Figure 9.3a. Each takes a measurement of the rate of spin of the approaching axle. Now there is a subtle surprise, each observer obtains the **same** result. Thus, if the axle approaching A is seen by A with a clockwise rotation, then observer B will also see the axle approaching them with clockwise rotation, Figure 9.3b. In this experiment both the initial explosion and the two subsequent observations have bilateral symmetry, a rotation of

180 degrees leaves the situation unchanged, except for my labels A and B. Therefore, we should expect the results of the two observations to be the same. It is important to focus only on the spin of the axles, because provided nothing interferes with the axles in their flight, the angular momentum of each axle will be conserved. Whatever the distance from the site of the explosion the spin measurement will be unchanged.

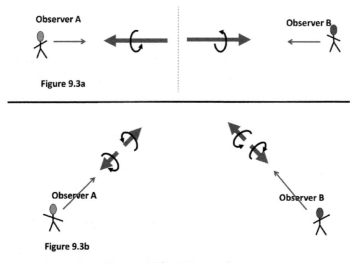

Figure 9.3: The EPR experiment

I now move from this classical illustration to the quantum experiment involving polarised light, or forms of matter, and consider this experiment using the systemic process approach. We again focus on the measurement of spin and a quantum system is created which has bilateral symmetry due to the experimental setup. Whilst we have no information about the created system, at a subsequent time that system is tested by two observers A and B at two perspectives at a distance from the site of the original experiment which also have that same bilateral symmetry. From the considerations of the symmetry we should expect the results of the two observers' experiments to be identical. We cannot conclude by analogy with the spinning axle case that two particles have been identified with equal and opposite spins. In the quantum case all that can be concluded when we subsequently compare the results of the two experiments is that the second experiment confirms the result of the first. Nothing further needs to be postulated to explain the experimental results. There is no necessity to propose superluminary communication between the A and B which has given so much cause for concern when formulated in the object paradigm.

Symmetry, inherent in many situations under scrutiny, plays a guiding role in choosing an appropriate modelling basis. If you enjoy, as I do, exploring an atlas, you will know to be careful in the way in which you interpret a map of the world. To map the world, the spherical earth must be flattened so that it may be drawn on a page. This inevitably distorts the relationships between different parts of the map and reflects the problems of using coordinates appropriate to a flat page to describe geometric shapes on the curved surface of a sphere. It can be done but leads to great complexity and difficulties in understanding. Whereas using an appropriate basis facilitates understanding, that is putting the map on a globe and using latitude and longitude or spherical coordinates as the basis for modelling on the surface of a sphere.

The natural world is a fractal domain, therefore the modelling basis for describing natural systems should itself be fractal to facilitate understanding. Therefore, I propose that the systemic process approach developed in the previous chapters is a more appropriate foundation for understanding and modelling natural systems than the traditional WEIRD approaches, particularly in physics, using the concept of a particles in empty space. This concept is a construction as discussed in Chapter 7 and does not have the required symmetry for building models in a fractal systemic world.

For the quantum world, Bernard d'Espagnat observes "quantum field theory indicates atomism is basically wrong". And yet most extant attempts to model the quantum world start with the classical assumption that particles, fundamental at some level, underlie modelling approaches.

9.5.   Modelling Systemic Flows

> But the real glory of science is that we can find a way of
> thinking such that the law is evident.
>
> (Feynman, Leighton & Sands 1963)

Systems Thinking explores the nature of collective behaviours, the behaviours of interconnected systems. In the established way of thinking, Quantum Mechanics is the study of interacting particles, and their collective behaviour. The extant accepted approaches to Quantum Mechanics use ideas from Systems Thinking and do indeed contain systemic language and ideas. So, it does seem natural to use the general ideas of Systems Thinking in formulating the approach to Quantum Mechanics.

But in Chapter 2 I drew attention to a fundamental difference between the way in which systems were initially historically defined and modelled and the definition and modelling approach that I am using here. Systemic modelling was in the first instance based on the concept of 'object' in the accepted WEIRD way. The extant theories of Quantum Mechanics were built on that first historical definition of a system. The obvious question therefore now is – what are the implications and results of abandoning this and changing to the systemic process approach that has been developed in this book? What follows therefore is, as far as I am aware, a new way of understanding the mathematical models of Quantum Systems which leads to changes in approach.

Modelling in the discipline of physics uses mathematical models, so that in the construction and carrying out of experiments to produce observations for testing models, the results can be precisely measured.

Following on from the consideration of symmetry in the previous section, if I am to develop mathematical modelling which will enable exploration of the systemic world in an effective and efficient way, I must find a mathematical method of describing these systemic situations with a basis which is fractal.

The mathematical modelling of waves has this property, so not surprisingly it is this mathematics which is used to describe Quantum Mechanics aligning the fractal symmetry of the mathematical modelling with the fractal structure of the epistemic modelling. I note that Wave Mechanics has been an alternative label for Quantum Mechanics.

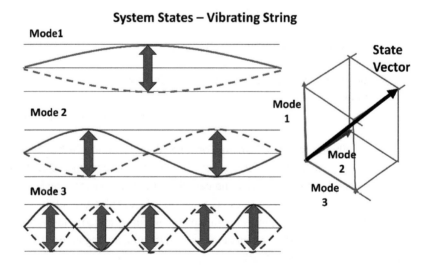

Figure 9.4: System states and state vector in classical physics

In general, mathematical models are based on the concept of an object. We count objects, we measure objects, we describe different types of objects in terms of their attributes. This is the WEIRD way of modelling and thinking. Quantum Mechanics traditionally describes a system by means of a vector which represents the internal state of the system under scrutiny. The vector is an object which has length and direction. This use is carried over from classical physics where the dimensions of the vector space in which the vector sits represent the attributes of the system, and the components of the vector represent the various values of the attributes. Recall the violin string in Chapter 3, where the modes of the string would be represented by the dimensions of the space, and the lengths of the components the amplitudes of each of the modes. Three such modes are illustrated in Figure 9.4.

But in Quantum Mechanics things get rather more complex because each value of what was the amplitude is now recognised as a different state. So if we just think of one string of the violin there is, in principle, an infinity of possible states, giving rise to an infinity of possible dimensions of the vector space. It gets a bit more technical with the introduction of complex numbers and Hilbert Spaces but still the dimensions of the space represent the states of the system just as before. But now the length of each component of the state vector represents the probability of finding the system under scrutiny in that state. The state vector now represents a set of probabilities for a particular combination of states.

Thus, in quantum models the length of any particular component of the vector indicates the probability of obtaining a particular value in an experiment to measure that value. In both situations the vector represents the state of the object, violin or atom, in the situation being described. The description is of that object, violin string or atom, as if there was no connection between the object and its environment. From a systemic perspective this has serious flaws: it does not carry the understanding that violins, and atoms never stand alone. Having understood that Quantum Mechanics is a process-based theory, this description is clearly inadequate because the vector description carries no intimation of process. The only dynamic is the changing internal state of the system; the ongoing flow of the environment is missing, as is the connection between the system and its environment. In order to overcome this, the concept of quantum entanglement was proposed, adding into the model a connection between objects under consideration which leads to further complexity. Now, to simplify matters I return to the flowing world of Heraclitus. To model flows I need to turn to the work of Ilya Prigogine (Kondepudi & Prigogine 1998). A flow can be described mathematically by a vector. Whilst any vector is of course an object, it can represent a flow: it represents the flow in magnitude and direction. A system of interest will have a number of input flows and a number of output flows. These flows will be independent of each other outside the system boundary but interacting within the system of interest. I construct a vector space as before but this time each dimension of the vector space represents either an input flow to the system or an output flow from the system. The value of each component in this representation is the magnitude of the flow. I now construct a state vector in the classical case exactly as was done previously by the addition of the components, but with one crucial and important variation. I add the components of the output flow and the negative of the components of the input flow to arrive at the state vector. The mathematical vector representing the system as a whole is therefore constructed from the vector sum of the output flows minus the vector sum of the input flows. This construction is both necessary and sufficient to represent a system in its environment.

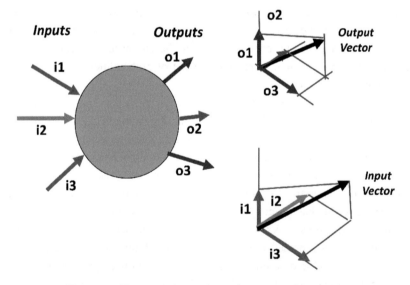

Figure 9.5: Vector representation of a system of interest

This representation has the required property that it is a fractal representation as would be required of any systemic representation. That is the resulting vector sum of vectors representing a group of interacting sub-systems making up a system does indeed represent the system itself. All internal flows sum to zero in the addition.

It remains to be seen whether this reinterpretation of what a state vector represents will provide a new approach to Quantum Modelling but it seems to me to have potential. The leap to probabilistic quantum modelling can be made in the same way as is done in the conventional approach.

First the sheer complexity of the representation of interacting systems has vanished. Previously, to build the vector space of two interacting systems it was necessary to construct that space as a tensor product of the two vector spaces. For every state of the first system there could be any state of the second, so that the dimension of the combined system was the product of the two individual systems. Taking as an example a violin and a cello, any note of the violin could be accompanied by any note of the cello. But in the real systemic duet the players and the feedback loop between the players must be included: hands $\rightarrow$ violin $\rightarrow$ sound $\rightarrow$ ears $\rightarrow$ hands $\rightarrow$ cello $\rightarrow$ sound, etc which produces the harmonies. Mathematically, in my alternative interpretation the interaction between the two systems takes place in the intersection of the two spaces representing each system flow. But for the whole system (that is, the system that results from the interaction of the two systems) the dimension of the space representing the whole system is the sum of the dimensions of the two spaces less the dimension of the intersection of those two spaces. The flows between the two interacting systems and the dimensions representing them play no part.

9.6.   Conclusions

In a comprehensive analysis, Bernard d'Espagnat (2006) concludes that Quantum Mechanics is a process theory and does not fit the realm of WEIRD thinking: a process being the fundamental entity on which the models are based. I have therefore concluded with Stafford Beer that Quantum Mechanics is a systems theory. Its foundation lies in the systemic process paradigm, not in the WEIRD thinking paradigm.

In the world of systemic modelling, time is present in the flows into and out of a system in a steady dynamic stationary state. Nevertheless, the system in focus is changing its environment because of those steady state flows. It is of course interacting with other systems. Changes to that steady state of the system in focus, as a result of interaction with other systems, will inevitably ensue. The understanding of the importance of perspective gives rise to a much restricted view of the importance of the results of any particular experiment and the need to identify and remove unwarranted assumptions.

But within process theory, as others have written (e.g. Griffin, D.R. (ed) 1986), time plays a fundamental role, and as Jean Piaget (1954) posited, spatial concepts are constructed by our brain and nervous system. This understanding reverses the dominant reductionist view of space as fundamental, time as a construction. Any observer's model of space is constructed from changes in perspective. The systemic process paradigm reverses the WEIRD scientific view of modelling that space is fundamental and time must be introduced, and takes the view that time is fundamental, because it underlies human perception.

Furthermore, a conclusion from accepting that Quantum Mechanics must be understood as a systemic process theory is that until there is a process theory of gravitation no progress can be made on unifying quantum theory and gravitation. Since the presently accepted theories are formulated in different paradigms, one a process or systemic theory, the other a traditional object-based theory, as they stand at present they cannot be brought together.

What is clear is that a systemic approach being aligned with the way in which Quantum Mechanics is approached dissolves problems which have caused much discussion and argument in the interpretation of the mathematical models. Although there remain many questions, it seems to me worth exploring further.

10. Looking Back

As we face a very uncertain future, the answer is not to do
better what we have done before. We have to do something
else. The challenge is not to fix this system but to change it; not
to *re*form it but to *trans*form it.

(Robinson 2015)

10.1 Introduction

The conclusion that really strikes me coming from the journey of exploration
that I have travelled is that the explanatory power of Systems Thinking is far
greater than I ever imagined. Starting from Heraclitus view that all is flux and
change (Chapter 4) forces the recognition that the natural environment for
any living form on planet earth is dynamic. Therefore any living form must
deal with the many threats and opportunities that this ever changing
environment brings. There are the threats and opportunities which are
routinely met, but there are also those which have never been seen before. In
both cases it is possible to identify common patterns of control and learning.
Common patterns that I will come back to later.

At the heart of this understanding is that there are two strands of Systems
Thinking and only when that differentiation becomes clear can progress be
made. The understanding that each of us has a unique perspective plays an
important role, and is central to the process of learning. To model a process, a
technique to represent that process in terms of a process model is needed, and
Western Science attempted this in the beginning in a very clumsy indirect
way, not up to the task. In this Conclusion I briefly explore why Beer's
approach, that I have termed *process systemic thinking,* can be considered a
distinct paradigm, and the implications of applying this thinking and
modelling techniques to the areas I have considered.

10.2 WEIRD Thinking is not the only way

The European culture of Science developed through the last 300 years or so
misses all that is contained in this book. We have been seduced by the success
that reductionism and realism have had in building understanding of the

relatively simple situations that were studied by the first scientists, Galilleo, Newton, Leibniz, and those that followed. The quest for 'objectivity' blinded natural scientists to the importance of perspective, something which is central to the understanding of social science, and therefore to understanding that the models we construct are just that, the human view of the human world.

In order to define further the relationship between WEIRD thinking and the approach I have taken I wish to make the distinction between *formal models* and *mental models*. Formal models are developed by scientists and others in order to understand and communicate aspects of our environment and ourselves. Mental models are those models contained in a brain and nervous system that enable an animal to remain viable. The discussion in Chapter 2 reaches the conclusion that the mental models which enable an animal to remain viable are patterns with the structure 'input→process→ output'. They are 'systems', whereas the formal models of WEIRD thinking are object-based models. Initially, all attempts to build formal models of processes by systems thinkers and others were within an object based 'scientific' approach, that is process models were constructed indirectly from an object basis. As noted in Chapter 1, and 7, Ernest Nagel (1961) sought to justify this approach. But then Stafford Beer (1979) developed formal models of processes without recourse to building from an 'object' basis, explicitly stating that these models were subjective and necessarily constructed from a particular perspective and characterised by a 'purpose'(Beer 1979).

In Chapters 6 and 7 I proposed that these formal *process systemic models* create a reasonable explanation for the difference between the WEIRD thinking of western societies and the modes of thinking that sustainable societies use. (see for example Yunkaporta 2019) But the hypothesis that the groups of people who think differently to WEIRD thinkers think systemically does need thorough investigation before it can be substantiated. However, it is my contention that we are born systems thinkers and only through education are we trained to think and communicate only in terms of objects. From my teaching experience I have learnt that there is a range in students from those who are natural systems thinkers and find no difficulties in adopting the modelling techniques described in this book, to those who reject Systems Thinking entirely. Therefore I find no surprise that Henrich, who with others, first published the idea that WEIRD Thinking is indeed weird in human cultural terms (Chapter 1), proposes that the thinking across different societies is multidimensional covering a continuum across different cultures (as reported in an interview published in *New Scientist* 5th September 2020).

Conceiving the world 'out there' in terms of formal systemic process models is fundamentally different from conceiving the world out there in terms of object models. Starting from a *systemic process* view, I have proposed in Chapter 7, that at a particular time a young animal views a scene and differentiates out from that scene a particular feature which, when the animal changes its perspective, holds some constancy. Over time the animal builds an *object* mental model of that feature in its environment. That mental model is a schema system synthesised from the many perspectives that it experienced of that feature. The order in which it happened to take the perspectives which gave rise to the object model is immaterial to the eventual complete object model. The object model is, therefore, time independent – its purpose is to enable the animal to recognise an object from any perspective. The process of constructing object models in this way is essential to survival in order to recognise predators and food whenever they appear in the animal's environment. Object models are inherently time independent but are used to predict the flow of external events from any perspective appertaining to a particular feature in an animal's environment. As a result, a brain can use the resultant object model from any perspective that occurs within a perceptual process as it unfolds. An object model is time independent, but a systemic process model is a model in which time is inherent in the flowing nature of the connections with its environment. The difference in the understanding of time and space between process systemic thinking and WEIRD thinking is therefore fundamental. Whilst the two modes of modelling are connected I conclude that they are inherently different paradigms, the object modelling paradigm derived from the systemic process paradigm as discussed in Chapters 7 and 9.

In science we have been misled by the attractiveness of object models independent of perspective. We miss that they are not only necessarily disconnected from their context, but also the detail of any particular perspective is lost. For normal living it is not necessary for a brain to retain all that detail, but only to recognise a situation moment to moment. The detail will be there, obtained from the situation as it is experienced. When constructing models an important further point, obtained from scientific model building, is that the conceptual basis from which it is built is best chosen to have the symmetries of the situation being modelled, as discussed in chapters 7 and 9. Object-based modelling certainly has its place, but attempting to use object modelling to model processes produces unnecessary complexity and creates problems of understanding. It was not until this became clear to me that I could begin to resolve the discord I experienced with published writing on Systems Thinking.

10.3 Understanding managing

As we navigate through a dynamic world our attempts to pursue our purposes are called managing. Managing an environment is central to surviving, and is by its nature a systemic activity. For every action by a system which changes its environment there is a reaction from that environment – a feedback loop. Understanding that feedback loop is central to managing. Perhaps the first hint that this was the case was Douglas McGregor's (1960) founding of the 'Human Relations Movement' (McGregor 1960) proposing his Theory X and Theory Y. He proposed that it was a manager's attitude to those managed that determined how those managed behaved. If a manager assumed the worst of those managed - that those managed are untrustworthy, lazy, and workshy, that is what they find – Theory X. On the other hand if a manager assumed the best - that those managed are trustworthy, enthusiastic, and hard-working, sure enough that is what *they* find – Theory Y. From a systemic perspective, with understanding of that feedback loop this is no surprise. It accords with my experience as a manager managing as Theory Y, and also from my experience of being in a group being managed when hierarchic control and Theory X prevails.

Embarking on a career in the teaching of management in 1994 it seemed natural to found that work in my own experience of managing and knowledge of Stafford Beer's systemic approach to management. But as I explored the literature on management and the wider literature on Systems Thinking I found that the literature on management was almost completely unaware of Systems Thinking, and it was far from accepting that there was any connection between Systems Thinking and managing. Beer is considered to be the founder of the discipline of Management Cybernetics, considered a sub branch of Management, but it became clear to me that Management is a sub branch of the science of Cybernetics, Chapters 4 and 5, that of controlling a human organisation. The effect of feedback in communication between individuals is dominant in human relations. Bregman (2020) explores the full extent that understanding the effect of this feedback has in human societies.

My feelings of growing discord with the dominant WEIRD view were brought to a head in 1998, when I began to teach the various techniques of Systems Thinking in a course for students pursuing a course as part of the study for a Masters degree in Business Administration. The literature on systems thinking contains many perspectives and attempts had been made to relate the different techniques (See for example Jackson 2003) but once I was able to define the difference between WEIRD thinking and Beer's systemic process thinking I was able to relate the various strands in a satisfactory way. The journey described in this book offers my conclusions from that exploration. These conclusions are of course for others to explore and either verify or reject as scientific method requires.

Applying the lessons from Chapters 3 and 4, as in all human organisations, it is often the case that those in management positions have not grasped the fact of the complexity that faces them. It is only because those around them agree to being managed, that the organisation is able to function. It is cooperation, not bullying, not competition, nor control, that is the principal glue that enables the organisation, or a society, to function. It has become clearly apparent now in 2020 that when there is no cooperation, as with attempts to manage a viral pandemic, a typical management which has been lulled into that false sense of security, simply fails, and the more centralised the decision making in that management hierarchy the worse the failure. Those supposedly managing do not have the requisite variety of potential command to succeed.

Beer's leap of understanding was to look to natural systems which do achieve adaptability and resilience, and then connect this with the cybernetic mathematical rules, the Law of Requisite Variety, the Conant-Ashby Theorem, and the Sub-optimisation Theorem (Beer 1966). He gained inspiration from the human body realising the necessity of a fractal structure as I have described in Chapters 4 and 5 (Beer 1971). Achieving the necessary variety of potential command, can only be achieved through a layered fractal structure and the rule set out in Figure 4.12. This rule is only achieved in a human organisation through discussion in a feedback loop between layers of management in that organisation. Once the idea of *semi*-autonomous sub-units within a framework agreed between two levels, is established, as Beer's VSM proposes, then the two levels can work together to the advantage of both. That framework is never fixed but must evolve to enable flexibility of both unit and sub-unit within its evolving environment. Only in this way can an organisation use all the available brainpower to its advantage. This is the secret of adaptability and resilience for managing an organisation, from the restaurant considered in Chapter 4, to the community in Chapter 9.

10.4 Learning

But to those conclusions I must add that learning is central to successful management. Problems come to an animal, an organisation or a society from many different directions and the capacity of any brain is limited. To be able to respond to not only issues known, but be ready for all the unknowns needs both constrained and open learning (Chapter 6). But if in an organisation or community learning is constrained by culture, by propaganda, by diktat, or law there will inevitably be a failure to manage at some point. As discussed in Chapter 5 there is always a balance to be struck between the framework of culture and law and the freedom of any organisation or community to find solutions to the problems it faces.

Once a culture becomes established it becomes more and more difficult to transcend the way of thinking that is imposed. The culture defines what is acceptable and what is not, only constrained learning is possible. Despite culture, despite propaganda, and instead of dismissing other ways of living and thinking as inferior, as those at the centre of any hierarchy always seem to do, I suspect many people *do* listen to others. It has always been possible to listen with kindness to those who live and think in different ways, but dominant male establishment hierarchies seem not to be geared to listen to challenges to their way of thinking. The thinking within a combination of a dominant ideology and a standard mammalian male hierarchy allowing only constrained learning does not change easily.

It is perhaps significant that in Ancient Greece the great advances in philosophy and science were made there because of the geography of the region, in that there were interconnected semi-autonomous sub-units of learning scattered around the Mediterranean Sea. Scholars travelling between centres and spending time in different centres would trigger synthesis of ideas; the open learning described in Chapter 6. These were then extinguished by the Roman centralised state. The next similar situation arose with the Islamic centres of learning, and following that those that emerged in the Renaissance. The disconnected nature of the sub-units allow the development of different cultures and ways of thinking, which when they do come together can make leaps of learning.

I conclude that interconnected networks of subunits which can develop their own ways of thinking but can come together in an open learning situation are necessary to allow the evolution of our human ability to manage our situation on this planet. We must listen with kindness to others from other cultures to have a chance of human survival.

The common pattern of semi-autonomous sub-units evolving but connected to others is the pattern of modular scale free networks (Barabasi 2002), a pattern common to the subsystems of Beer's VSM, and a pattern ubiquitous in nature. It is close to the way in which classes in primary schools in the UK are set up, but not secondary schools. The quotation at the head of this chapter was addressed to the UK's education systems. As Robinson writes, at present we are failing to engage children in learning even though their natural inclination is to be engaged in learning. If we are to improve self-management, management of enterprise, and improve government, then we should be redesigning secondary schools and include in the curriculum systems thinking, and its applications to raising children, government of communities, and managing our environment. These are aspects of life which affect all.

10.5 Governing – learning from nature

In our current world we spend inordinate amounts of time and effort in exploring 'improvements', so called, in arms, and warfare but very little indeed on exploring how we might improve our governing to be able to live together peacefully, sustain ourselves, and manage our environment sustainably. From a systemic perspective the question to ask is - 'What is the purpose of this enormous spending? More than anything it seems to be just pandering to the egos of rogue males. But, does it help sustainability to have robot armies facing each other when we are extinct? This seems the logical conclusion of our present path. In order to survive we must develop governing structures capable of fulfilling the purposes of maintaining peaceful order in human societies, and sustainably manage the ecosystems within which we live. None of our present governing structures have the requisite variety of potential command to achieve this. Few, if any, satisfy the three rules of management. The dominant economic ideology seeks optimisation of sub-units, and thus great disparities of income and wealth become inevitable. Dominant sub-units – countries, and corporations automatically evolve to take the income and wealth to themselves and we end up far from optimising for the benefit of all. We construct centralised governing hierarchies which can neither satisfy the Law of Requisite Variety nor the Conant Ashby Theorem, but there is another path, that of promoting trust and decentralisation, bringing back control to each and every person, within a fair and equitable framework.

Even in the mainstream of Western thinking on these problems it is recognised that current systems are failing. But Thomas Piketty's analysis of historical economic evolution of a number of societies (Piketty 2020) shows that it has been possible in the past to improve the situation within nations. The years from 1918 to 1970 saw great improvements in health and economic well-being across a number of countries but from around 1970 these improvements have subsequently been reversed. Wealth disparities which were reducing have increased, and the growth of economic well-being has also declined. Piketty states that we need to solve three problems to construct a successful governing system – a theory of borders, a theory of property, and a theory of deliberation (Piketty 2020 p630).

We must recognise that the level of complexity in managing a community is much higher than in managing an enterprise of the same size and therefore managing a community is much more difficult. Any governing system must follow the rules of management. As described in Chapter 8, Beers VSM gives the framework for a theory of borders the creation of semi-autonomous sub-units in a fractal structure. Whilst this is far from a simple change, in particular in an urban society, over time such a structure can be established. By so doing and allowing for the contact that was discussed in Chapter 8 aids learning.

The theory of discussion comes first from the necessity of ensuring that the bodies House of Representatives, House of Expertise, and Communities Collective are all structured appropriately to be able to achieve their purposes. Again it is will not be easy to insure that the House of Representatives does contain as far as possible all cultural perspectives, and that the House of Expertise is outwardly focused on the evolution of the technical world, the environmental world, and the geographical and cultural world. But it is essential that they do in order to become learning systems. The Communities collective must also be a learning system, focused in order that communities are able to learn from each other. Beer's Team Syntegrity (Beer 1994) offers further thoughts on a theory of discussion which follow the principles which enable open learning.

Piketty's last requirement is a sustainable theory of property, and again this is necessarily complex. As Piketty argues the extremist positions of no private property at all, as in the aspiration of Soviet Communism, or no public property as in the aspiration of American capitalism are both too simplistic and neither sustainable. As is argued in Chapter 8 any community of whatever size in normal times must be able to decide for itself, perhaps with others, its evolutionary path. In a sustainable world there can no devastation of communities, as seems to be common in the present world, where communities in both extremes have no capacity to influence decisions which impact them. At least for me the application of process systemic thinking begun in Chapter 8 offers a path towards a sustainable human civilisation and sustainable environment on planet earth.

10.6 Reformulating physics

When I met Quantum Theory in my degree studies as an undergraduate in 1965 for the first time it produced a discord, the mathematical techniques used to explain the experimental findings did not seem satisfactory. This feeling has persisted right through to this present day even despite the research I undertook and teaching undergraduates Quantum Mechanics myself in 1972. Even though teaching forces better understanding, the better understanding achieved did not dissipate the discord. If process systemic thinking does underpin WEIRD thinking as I now believe then Quantum Theory is indeed the fundamental basis of physics. My expectation is that thinking through how the change to the process systemic understanding of a system will give new possibilities the very first beginnings of which I have tried to set out in Chapter 9. I hope others take up the exploration along this path.

Bibliography

Ampère, A.E. (1834) *Essai sur la Philosophie des Sciences*, Bachelier

Angyal, A. (1965) *Neurosis and Treatment*, John Wiley

Aristotle (1986) *De Anima*, trans H. Lawson-Tancred, Penguin Books

Arnold, C. (2013) 'The Other You', *New Scientist* 12th January

Ashby, W. R. (1952) *Design for a Brain* [revised ed, 1960, Chapman and Hall]

_____ (1956) *An Introduction to Cybernetics*, Chapman and Hall

Asby, R. (1969) 'Optical-Mode Interaction in Nonlinear Media', *Phys. Rev.* 187, 1062

_____ (1969a) 'Theory of Resonant Optical Second-Harmonic Generation from a Focused Gaussian Beam', *Phys. Rev.* 187, 1070

_____ (1969b) 'On the Theory of Second Harmonic Generation', *Journal of Opto-Electronics* I, 165-171

_____ (1970) 'Theory of Optical Parametric Amplification from a Focused Gaussian Beam', *Phys. Rev. B* 2, 4273

_____ (1973) 'On the Theory of Radiating Electrons', in *Coherence and Quantum Optics*, Eds L. Mandel and E. Wolf, Plenum Press

_____ (2010) 'System Dynamics', in *TU811 Thinking Strategically: Systems tools for managing change*, Open University Press

Ball, P. (2018) *Beyond Weird: Why everything you thought you knew about quantum physics is different,* University of Chicago Press

Barabási, A. (2003) *Linked: The New Science of Networks*, Plume

Bateson, G. (1972) *Steps to an Ecology of Mind*, Ballantine Books

_____ (1979/2002) *Mind and Nature*, Hampton Press

Bayer, H. C. von (2016) *QBism: The Future of Quantum Physics*, Harvard University Press

Beer, S. (1959) *Cybernetics and Management*, English Universities Press

_____ (1966) *Decision and Control*, John Wiley

_____ (1972) *Brain of the Firm*, [2nd ed, 1981, John Wiley]

_____ (1974) *Designing Freedom,* Garden City Press

_____ (1975) *Platform for Change*, John Wiley

_____ (1979) *The Heart of Enterprise*, John Wiley

_____ (1985) *Diagnosing the System for Organizations*, John Wiley

_____ (1994) *Beyond Dispute: The invention of team syntegrity*, Wiley

Bernard, C. (1957) *An Introduction to the Study of Experimental Medicine*, Macmillan

Bertallanfy, L. von (1968) *General System Theory: Foundations, Development, Applications* [revised ed, 1976, George Braziller]

Biggart J., Dudley P. & King F. (1998) *Alexander Bogdanov and the Origins of Systems Thinking in Russia*, Ashgate

Boring, E. G. (1930) 'A New Ambiguous Figure', *American Journal of Psychology*, vol. 42, p. 444

Bregman, R. (2020) *Humankind: A hopeful history*, Bloomsbury

Bullock, A. and Stallybrass, O, (1977) *The Fontana Dictionary of Modern Thought*, Fontana Books

Cannon, W.B. (2nd ed, 1939) *The Wisdom of the Body*, Norton

Carson, R. (1962) *Silent Spring*, Houghton Mifflin

Checkland, P. (1981) *Systems Thinking Systems Practice* [revised ed, 1999, John Wiley]

Cilliers, P. (1998) *Complexity and Postmodernism: Understanding complex systems*, Routledge

Conant R.C. and Ashby W.R (1970) 'Every Good Regulator of a System Must be a Model of that System', *Int J. Systems Sci*, Vol. 1 No. 2 pp.89-97

Craik, K. J.W. (1967) *The Nature of Explanation*, Cambridge University Press

Dahaene, S. (2020) *How We Learn: The New Science of Education and the Brain*, Allen Lane

Damasio, A. (2006) *Descartes' Error*, Vintage Books

Davison, C.J. and Germer, L.H. (1928) 'Reflections of Electrons by a Crystal of Nickel', *Proceedings of the National Academy of Sciences of the United States of America*. 14 (4): 317–322

Dawkins, R. (1989) *The Selfish Gene*, Oxford University Press

Deely, J. N. (2001) *Four Ages of Understanding*, University of Toronto Press

Denes, G. (2016) *Neural Plasticity Across the Lifespan*, Routledge

d'Espagnat, B (1999) *Conceptual Foundations of Quantum Mechanics* (2nd ed) Perseus Books

_____ (2003) *Veiled Reality: An Analysis of Present-Day Quantum Mechanical Concepts*, Westview Press

_____ (2006) *On Physics and Philosophy*, Princeton University Press

Dewey, J. (1938) *Experience and Education*, Kappa Delta Pi

Diamond, J. (2005) *Collapse: How Societies Choose to Fail or Survive*, Allen Lane

_____ (2013) *The World Until Yesterday: What can we learn from traditional societies?* Penguin Books

Eagleton, T. (2011) *Why Marx Was Right*, Yale University Press

Edelman, G.M. (1987) *Neural Darwinism: The theory of neuronal group selection*, Basic Books

Ewing, J.A. (1899) *The Steam Engine and Other Heat Engines*, Cambridge University Press

Feferman, S, et al (Eds) (1986) *Kurt Gödel: Collected Works* (Vol I), Oxford University Press

Feynman, R.P., Leighton, R.B. & Sands, M. (1963) *The Feynman Lectures on Physics* (Vol I) Addison Wesley

Forrester, J. W. (1971) *World Dynamics*, Wright-Allen Press

Frisch, Karl von (1967) *The Dance Language and Orientation of Bees,* Harvard University Press

Getner, D. & Stevens A.L. (1983) *Mental Models*, Psychology Press

Glasersfeld, E. von. (1995) *Radical Constructivism: A way of learning,* Routledge

Gregory, R.L. (2009) *Seeing Through Illusions*, Oxford University Press

Griffin, D.R. (ed) (1986) *Physics and the Ultimate Significance of Time*, State University of New York Press

Hebb, D.O. (1949) *Organization of Behaviour: A Neuropsychological Theory*, Psychological Press

Heisenberg, W. (1958) *Physics and Philosophy*, Harper Collins

Helmholtz, H von (1866) *Handbuch der Physiologischen Optik*, English trans (1924) by J.P.C.Southall, *Treatise on Physiological Optics Volume III*, Dover Publications

Henrich, J., Heine, S. & Norenzayan, A. (2010) 'The Weirdest People in the World?', *Behavioural and Brain Sciences* 33:61-135

Hobsbawm, E.J., (2nd ed, 1990) *Nations and Nationalism Since 1780,* Cambridge University Press

Hofstadter, D. (1979) *Gödel, Escher, and Bach: The Eternal Golden Braid*, Basic Books

Ison, R. L. & Straw, E. (2020) *The Hidden Power of Systems Thinking: Governance in a climate emergency,* Routledge

Jackson, M.C. (2003) *Systems Thinking: Creative Holism for Managers,* Wiley

Jackson, T. & Webster, R. (2016) *Limits Revisited: A review of the limits to growth debate,* Report to the All Party Parliamentary Group (APPG) on limits to growth, http://limits2growth.org.uk/revisited

James, W. (1890) *The Principles of Psychology,* Holt

_____ (1907/1981) *Pragmatism,* Hackett Publishing

Johnson-Laird, P. N. (1983) *Mental Models,* Cambridge University Press

Kahn, Charles H. (1979) *The Art and Thought of Heraclitus,* Cambridge University Press

Kheirbek, M.A. and Hen, R. (2014) 'Neuroscience: Add Neurons, Subtract Anxiety', *Scientific American,* Vol 311, No 1, July

Kirk, G.S., Raven, J.E. & Schofield, M. (2nd ed, 1999) *The Presocratic Philosophers,* Cambridge University Press

Koestler, A. (1970) *The Act of Creation,* Pan Books

Kolb, D.A. (1984) *Experiential Learning: Experience as the Source of Learning and Development,* Prentice-Hall

Kondepudi, D. & Prigogine, I. (1998) *Modern Thermodynamics: From Heat Engines to Dissipative Structures,* John Wiley

Kuhn, T.S. (1996) *The Structure of Scientific Revolutions,* University of Chicago Press

Lakoff, G. (2009) *The Political Mind,* Penguin Books

Lettvin, J.Y.,Maturana, H.R., McCulloch, W.S. & Pitts, W.H. (1959) 'What the Frog's Eye tells the Frog's Brain', *Proceedings of the Institute for Radio Engineers,* 47 (11) 1940-1959

Levitt, M (2019) *Why Did Ancient States Collapse?* Archaeopress Publishing

Luard, N. (1981) *The Last Wilderness: A Journey Across the Great Kalahari Desert,* Simon and Schuster

McCulloch, W.S. (1965) *Embodiments of Mind,* MIT Press

Mandelbrot, B.B. (1982) *The Fractal Geometry of Nature,* Times Books

Martinez, R. (2016) *Creating Freedom: Power, Control, and the Fight for our Future,* Canongate Books

Maturana, H.R. and Varela, F.J. (1980) *Autopoiesis and Cognition,* D Reidel

_____ (1998) *The Tree of Knowledge,* Shambala Publications

Mead, C.A. (2000) *Collective Electrodynamics: Quantum Foundations of electromagnetism*, MIT Press

Meadows, D.H., Meadows, D.L., Randers, J. & Behrens, W.W. III (1972) *The Limits to Growth*, Earth Island

Medina, E. (2011) *Cybernetic Revolutionaries: Technology and Politics in Allende's Chile*, MIT Press

Mill, J.S. (1989) *On Liberty and other writings*, Cambridge University Press

Morecroft, J. (2007) *Strategic Modelling and Business Dynamics: a feedback systems approach*, John Wiley

Mukherjee, S. (2016) *The Gene: An Intimate History*, The Bodley Head

Nagel, E. (1961) *The Logic of Scientific Thought*, Routledge, Keegan, and Paul

Nisbett, R. E. (2005) *The Geography of Thought*, Nicholas Brealey

Ostrom, O. (1990) *Governing the Commons: The evolution of institutions for collective action*, Cambridge University Press

Peirce, C.S. (1878) 'How to Make Our Ideas Clear', *Popular Science Monthly* 12, 286-302

Perry, W.G. (1999) *Forms of Intellectual Development in the College Years; a scheme*, Jossey-Bass

Pert, C.B. (1997) *Molecules of Emotion*, Simon and Schuster

Piaget, J. (1954) *The Construction of Reality in the Child*, Basic Books

Picketty, T. (2020) *Capital and Ideology*, Harvard University Press

Plath, S. (1966) *The Bell Jar*, Faber and Faber

Plato (1894/1986) *The Republic*, trans Benjamin Jowett, Prometheus Books

_____ (1987) *The Republic*, trans Sir Desmond Lee, Penguin Classics

_____ (1994) *The Republic*, trans Robin Waterfield, Oxford World Classics

_____ (2006) *Alcibiades I and II*, trans Benjamin Jowett, The Echo Library

Popper, K.R. (1963) *Conjectures and Refutations: The growth of scientific knowledge*, Routledge & Kegan Paul

Ribeaux, S.E. & Poppleton P. (1978) *Psychology and Work: An Introduction*, Macmillan

Roberts, N., Anderson, D., Deal, R., Garet, M. & Shaffer, W. (1997) 'Introduction to Computer Simulation – A System Dynamics Modeling Approach', *Journal of the Operational Research Society*, 48:11, 1145

Robinson, K. (2015) *Creative Schools*, Penguin Books

Schrödinger, E. (1958) *Mind and Matter*, Cambridge University Press

Senge, P. (1992) *The Fifth Discipline - The Art and Practice of the Learning Organization*, Century Business

Shaffer, W.A. (1983) *Introduction to Computer Simulation: A Stems Dynamics Modelling Approach,* Addison Wesley

Shannon, C.E. & Weaver, W (1949) *The Mathematical Theory of Communication*, University of Illinois Press

Sherburne, D.W. (1966) *A Key to Whitehead's Process and Reality,* University of Chicago Press

Sherrington, C.S. (1920) *The Integrative Action of the Nervous System*, Yale University Press

Sinha, P. (2013) 'Once Blind and Now They See', *Scientific American* July, Volume 309, No 1

Thayer, H.S. (ed.) (1982) *Pragmatism: The Classic Writings*, Hackett Publishing

Vickers, G. (1972) *Freedom in a Rocking Boat*, Penguin Books

Von Foerster, H., ed (1951) *Transactions Eighth Conference*, Josiah Macy Jnr. Foundation

Werner, G. (Undated) *Fractals in the Nervous System: conceptual implications for Theoretical Neuroscience*, https://arxiv.org/pdf/0910.2741.pdf

Whitehead, A.N. (1929) *The Aims of Education*, Macmillan

_____ (1978) *Process and Reality*, The Free Press

Wiener, N. (1961) *Cybernetics*, MIT Press and John Wiley

Yunkaporta, T. (2019) *Sand Talk*, The Text Publishing Company

Index

About the Author

Robin Asby has spent a lifetime in problem solving. He is a systems thinker and researcher, a retired consultant and academic.

He started his career with a degree in Mathematics from Cambridge University (1965), and a PhD from London University in 1968 in Mathematical Physics which formulated the basic modelling of a focused laser beam. Subsequently he moved to the University of Rochester, NY and began research in optics and electrodynamics. His experience of a new culture, and the growth of interest in ecology sparked a new interest in systems thinking, politics and government. As a result of his experiences in the USA, when he returned to the United Kingdom in 1970, he became a political activist in the Labour Party. Moving to West Wales he spent 16 years training teachers and became familiar with the theory and practice of learning.

In the early 1980s he came across the systems thinking of Stafford Beer, and introduced systems thinking into his teaching. From 1989 to 1994 he established and ran a small enterprise working in the growing Information technology sector. Finding that West Wales was not fertile ground for such an enterprise, and that the problems of business were in the large majority problems of management, in 1994 he co-founded a new business department in a higher education college with courses designed from a systemic standpoint.

From 1998 to 2006 he worked as a consultant in conjunction with Hull University mentoring senior and middle managers in the Middle East, Far

East, and UK as they developed their understanding and use of systemic management theories in their own organisations. In 2001 he became an Open University tutor, then a contributor to the development of systems thinking courses. He chaired the development of the Open University MSc 'Systems Thinking in Practice'. He was a joint author of one of the courses in this degree.

He has a continuing interest in education. From 2000-2015 he was a governor of two secondary schools. In 2018 he was a founder member of Todmorden Learning Centre and Community Hub (tlchub.org.uk). TLCCH expects to establish a learning centre for research, development and practical learning in the former Todmorden Community College for those interested in building climate change resilient communities.

About the Publisher

Triarchy Press is an independent publisher of books that bring a wider, systemic or contextual approach to many different areas of life, including:

Government, Education, Health and other public services

Ecology, Sustainability and Regenerative Cultures

Leading and Managing Organisations

Psychotherapy and Arts and other Expressive Therapies

The Money System

Walking, Psychogeography and Mythogeography

Movement and Somatics

Innovation

The Future and Future Studies

In particular, please see:

Systems Thinking in the Public Sector (John Seddon)

Systems Thinking for Curious Managers (Russ Ackoff)

Growing Wings on the Way (Rosalind Armson)

Systemic Leadership Toolkit (William Tate)

For these and other books by Russell Ackoff, Barry Oshry, John Seddon, Rosalind Armson, Nora Bateson, Daniel Wahl, Bernard Lietaer, Phil Smith, Bill Tate, Sandra Reeve, Graham Leicester, Alyson Hallett and other remarkable writers, please visit:

www.triarchypress.net